GABY KENNARD'S WORLD FLIGHT

21 Dakar	**26** Cairo	**31** Madras	**36** Alice Springs
22 Agadir	**27** Luxor	**32** Phuket	**37** Parkes
23 Alger	**28** Bahrain	**33** Singapore	**38** Camden
24 Tunis	**29** Karachi	**34** Bali	**39** Sydney
25 Iraklion	**30** Bombay	**35** Darwin	

SOLO
WOMAN

Gaby Kennard's World Flight

To Dear Nonna,
Thank you for your
quiet support. It was
very important to me,
as particularly in the
early stages I had very
few.
with much love,
Gaby.

SOLO WOMAN
Gaby Kennard's World Flight

BY GABY KENNARD

SALLY MILNER PUBLISHING

BANTAM BOOKS
SYDNEY • AUCKLAND • TORONTO • NEW YORK • LONDON

SOLO WOMAN
A BANTAM BOOK

Printing History

Milner and Bantam edition published 1990
First published in Australia by Sally Milner
Publishing and Transworld Publishers

National Library of Australia
Cataloguing-in-Publication entry

Kennard, Gaby,
  Solo Woman

  ISBN 1 86359 008 0

  1. Kennard, Gaby. 2. Women air pilots —
  Australia — Biography. 3. Flights around the
  world. I. Title.

629.13092

Published in Australia by
Sally Milner Publishing Pty Ltd and
Transworld Publishers (Australia) Pty Limited,
15-25 Helles Ave, Moorebank NSW 2170
and in New Zealand by
Sally Milner Publishing Pty Ltd and
Transworld Publishers (NZ) Limited,
Cnr Moselle and Waipareira Aves,
Henderson, Auckland

Printed in Australia by
Southwood Press, New South Wales
Typeset by
Asset Typesetting Pty Ltd, New South Wales
Cover design by Doric Order
Text design by Doric Order

Text prepared with the assistance
of Kerry McAloon

I dedicate this book to my mother, Patricia Marguerite Buchan who, through her death, showed me what courage really is.

I acknowledge the inspiration of Amelia Earhart who willingly paid the price of courage.

ACKNOWLEDGEMENTS

I wish to thank the following people, without whose assistance my flight would have been impossible:

Hope Antzoulatas
The people of Atchison,
    especially Joe Carrigan,
    Dave Dennis, Virgene and
    Vince Smolik
The Australian Women
    Pilots
John Baxter
David Bell and family
Lynora Brooke
Paul Cope
Jan Coulter
Iris Critchell
The people of Darwin
John Death
David Dent
Saudamini Deshmukh
George Dukats
Alex Ewanchew
Barbara Evans
Sheri Forbes
Kathryn Flynn
Holly Friedman
Christine Gee
John Gilchrist
Margaret and Nort Gill
Fay Gillis-Wells
Philip Goard
Bob Grant
Michael Green
Captain Les Hayward
Aminta Hennessy
Neville Hill
Shirley Hourigan
Marcia Hremevuic
Daryl Jordan
Neville Kennard
Theo Kennard
Katherine Lien
Lois Luehring and Dave

George and Joanie Mallaliou
Geoff McLaughlin
Nancy Miller
Jack Munroe
Andris Murnieks
My neighbours —
    The Hamiltons
    The Nialls
Cliff Neville
The Ninety-Nines
Marcel Noe
Peter Norvill
Mary O'Brien
Michael Quinn
Eleanore Reichenbach
Bob Rice
Clive Roberts
Royal Aero Club Formation
    Fliers
Royal Australian Air Force —
    Warrant Officer
    Tony Tobler
    Flight Lieutenant
    Mark Thompson
Mohini Schroff
Ram, Sunita and
    Leila Shahaney
Peter Shilton
Rob Simpson
Dick Smith
Sherry Stumm
Kerry Taylor
Tony Vacarella
Nancy-Bird Walton
Pat Wilson
Col Whyte
Amanda and Doug Wright
Frank Young

SPONSORS

I wish to thank the following sponsors who had such faith in me:

AWA Electronic Services

A M R Combs

Aircraft and Component Engineering Pty Ltd

Australian Avionics

Australian Geographic

Bose Corporation

Channel 9 News and *60 Minutes*

Civil Aviation Authority

Ella Bache

Hardy Brothers Jewellers

Heath Fielding Group and G M Forsaith Aviation Pty Ltd

J P Cordukes Pty Ltd

Kennards Hire

McIntyre Marine Services Pty Ltd

Navair — for help regarding technical advice and overseeing all work on the aircraft

OTC Skycoms

QANTAS — for help in planning, operational matters, all overflight and landing clearances, moral support

Slender You South Pacific Pty Ltd

Simpson Aeroelectrics Instruments

CONTENTS

LIST OF PHOTOGRAPHS

Departure day, 3 August 1989, and it's time to say goodbye to my children — we're all being brave!

Just prior to departure, 3 August 1989, in my crowded cockpit, surrounded by fuel tanks, my maps and flight charts beside me.

My crowded cockpit from the back door, showing fuel tanks and luggage.   COURTESY AAP

Departing Bankstown, 3 August 1989.

August 1989, Lae, Papua New Guinea. The Papua New Guinea Air Force commanding officer and visiting RAAF officers pose after Amanda Wright, Ella Birrel and I had laid a wreath in honour of Amelia Earhart and Fred Noonan.

Lae, Papua New Guinea, and the strip Amelia Earhart last took off from.

The commanding officer took this picture of me as I was about to leave Johnston Island. It looks very touristy — what a false impression!

Motorcade through the streets of Atchison to the Forest of Friendship, with Virgene Smolik driving.
COURTESY DONALD E. MARTIN.

Dave Dennis, the Mayor of Atchison, Kansas, presents me with two keys to the city on my arrival, and makes a proclamation of Gaby Kennard Day, 5 September 1989.
COURTESY DONALD E. MARTIN

Amelia Earhart's birthplace, Atchison, Kansas.

The 99s and the people of Atchison commemorate my flight with this plaque in my honour. I was hoping, for their sakes, as well as for my own, that I would finish my journey.
COURTESY DONALD E. MARTIN

In the Forest of Friendship, Atchison, Kansas, 5 September 1989. Side by side with Amelia.
COURTESY DONALD E. MARTIN

Checking the oil, Memphis, Tennessee, 9 August 1989.
COURTESY NANCY MILLER

The Indian Women Pilots present me with a silver plate commemorating my solo around the world flight, in Bombay.

On Australian soil again. 4 November 1989, and I sign autographs for children in Darwin. The people of Darwin gave me a wonderful welcome.
COURTESY NORTHERN TERRITORY NEWS

Nearly home. Taxiing in for a briefing with the Royal Aero Club formation fliers, Camden, 10 November, 1989.

The Royal Aero Club formation fliers, with me, over Bankstown Airport, 10 November 1989.
COURTESY BURNIE PHILLIPS

About to touch down on runway 29er right, Bankstown, 10 November 1989.
COURTESY BURNIE PHILLIPS

Taxiing in at Bankstown, 10 November 1989. I'm home and all those people are waiting to greet me in the rain.

Hugging James and walking through the crowd at Bankstown, 10 November 1989. How wonderful to be home.
COURTESY BURNIE PHILLIPS

Starting on the motorcade through Sydney, with Mimi and James, 10 November 1989.
COURTESY BOB LIVINGSTONE

Excitedly arriving at Sydney Town Hall, surrounded by well-wishers. This was such an emotional time. These people are really here for me. How amazing!
COURTESY BOB LIVINGSTONE

Those wonderful members of the Australian Women Pilots, who helped and supported me all the way, at the Mayoral reception, 10 November 1989. On my left is Nancy-Bird Walton.
COURTESY BOB LIVINGSTONE

Together again with my children.
COURTESY TELEGRAPH

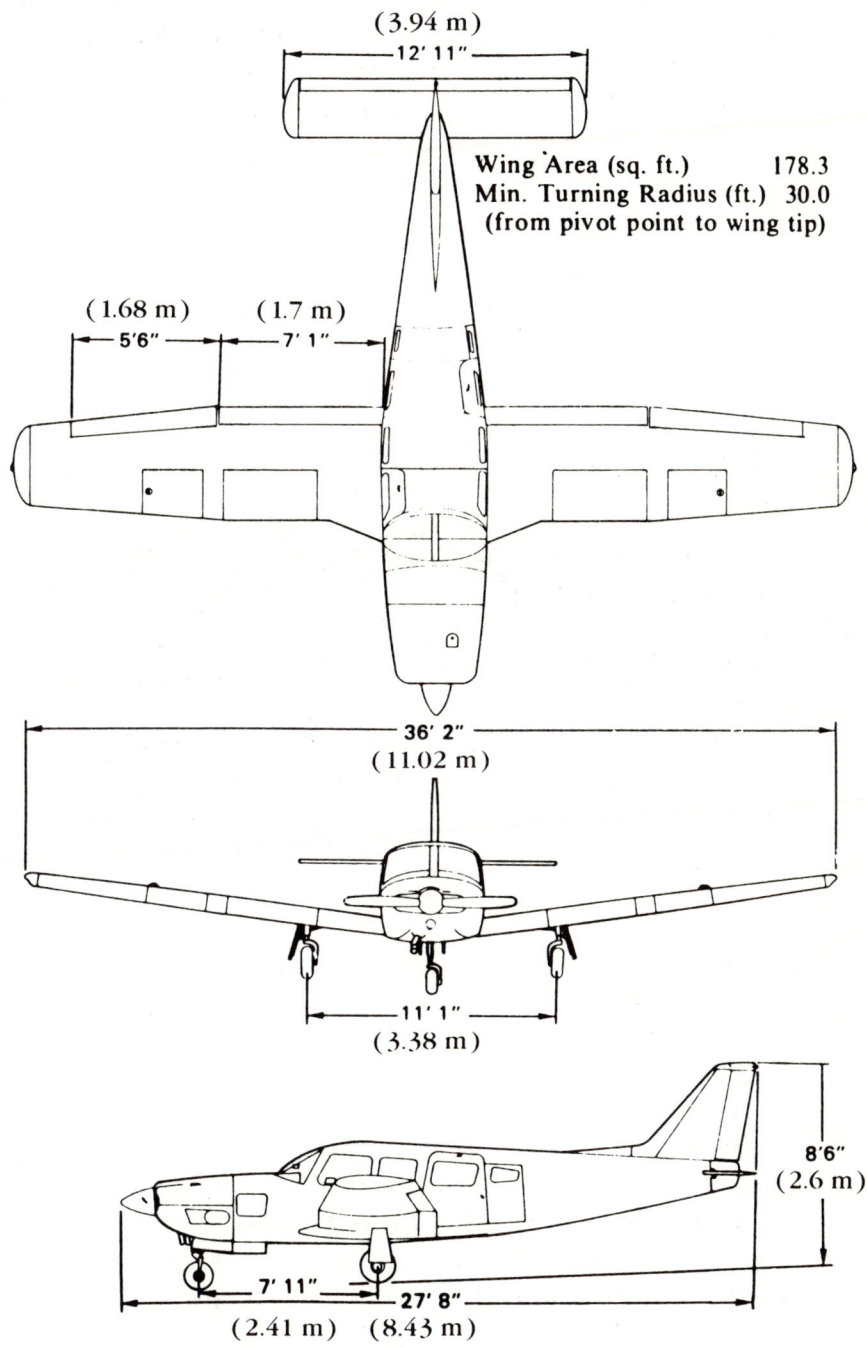

(3.94 m)
12' 11"

Wing Area (sq. ft.)   178.3
Min. Turning Radius (ft.)   30.0
(from pivot point to wing tip)

(1.68 m)   (1.7 m)
5'6"   7' 1"

36' 2"
(11.02 m)

11' 1"
(3.38 m)

8'6"
(2.6 m)

7' 11"   27' 8"
(2.41 m)   (8.43 m)

Plane Dimensions PA-32R-301 Saratoga

The diagram on these pages shows my instrument panel. I have tried, as simply as possible, to explain what all the items are, and I hope that these explanations will help the reader to understand the terms used in the book.

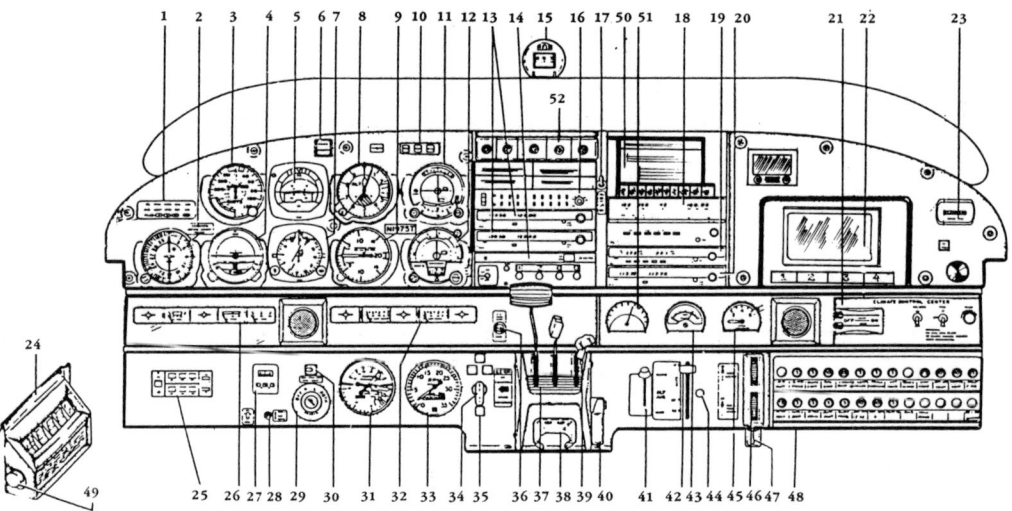

Instrument Panel

1 Autopilot mode annunciator — I can choose several modes, eg, heading only, altitude only, and so on
2 ADF Automatic Direction Finder — this picks up the signal from ground stations, and the needle points to the station — range is mostly around 100 nautical miles.
3 Airspeed indicator — indicated airspeed
4 Turn and bank — this tells me whether the plane's wings are level, or if the plane is in a slip or a skid

5   Flight attitude indicator — this gives an artificial horizon for use if I am flying on instruments and unable to see the horizon

6   Gear unsafe light — tells if the wheels are in transit — are not locked in the up or down position

7   DG Direction Gyro — a compass which is set from the main compass, and readily shows my direction of travel — it needs constant checking and resetting.

8   Altimeter — shows the height or altitude in feet

9   VSI Vertical speed indicator — shows the speed of ascent or descent

10   Annunciator panel — these are indicator lights to show if certain systems are working

11   HSI Horizontal Situation Indicator — this is my slaved compass, which automatically adjusts itself to the main compass, and which is connected to the VOR, and shows glide slope, localiser, allowing me to make an instrument landing

12 VOR indicator — very high frequency omnidirectional range indicator — this is a navigation instrument which enables me to know where I am and can tell the bearing from the ground station when in range

13   VHF radios 1 and 2 — very high-frequency radios (in 'line of sight' only)

14   Transponder — my transponder signal can be picked up on the ground if there is radar and tells the controller my position and altitude

15   Magnetic Compass — the primary navigation instrument

16   Audio selector panel — switches for the radio

17   Radio master switch

18   DME Distance Measuring Equipment — I can use this where there is a ground station giving off a signal and it tells me my distance from the station

19   ADF frequency selector — here I put in frequencies for the Automatic Direction Finder to pick up signals from ground stations which are called **Non-Directional Beacons (NDBs)**

20   Nav 1 and 2 radios — here I put in the frequencies for my two VORs

21   Climate control centre — I can adjust the air vents and heater

22   TV monitor — (extra for this flight)

23   Engine hour meter — how many hours I have flown

24   Master switch for the electrics and alternator, auxiliary fuel pump, anti-collision beacons, landing lights and pito heat (the pito tube senses the speed of the aircraft and if it ices up it is impossible to know the indicated speed of the aircraft)

25   Autopilot mode selectors

26   Engine instrument cluster — oil pressure, temperature, cylinder head temperature, alternator switches

27   Slavemeter — I can slave or unslave the HSI

28   Nav selector switch — I can choose which radios I use

29   Magneto and starter switch — ignition

30   Pitch trim switch — an electric system to operate the trim — I don't use this; I prefer to do it manually, with the trim wheel located between the pilot's and co-pilot's seat

31   Manifold pressure and fuel flow gauge — tells the fuel/air mixture, and how many US gallons per hour are being used

32   Fuel quantity gauges — show the main tanks only

33   Tachometer — revolutions of the propellor per minute. I have a variable pitch propellor

34   Gear selector switch — to put the wheels up or down

35   'Auto-Ext-Off' light — indicates if the emergency override system is on. Below a certain speed the wheels come down, but I could override this. At this speed a warning also sounds if the wheels are not down

36   Throttle

37   Propellor pitch control — adjusts the 'angle of attack' of the blades

38   Microphone

39   Mixture control — this controls the ratio of fuel to air

40   Friction nut — tightens the throttle so it can't be moved accidentally

41   Alternate air control — in the event of the air intake and filter becoming blocked by ice or some obstructions, I can use another air source

42   Wing flap selector — I put the flaps down in 3 stages to change the camber of the wing — it increases lift when flying slowly

43   EGT Exhaust Gas Temperature gauge — this is used to help determine the air/fuel ratio. The leaner the mixture, the higher the EGT

44   Flap in transit light — tells if the flaps are moving up or down

45   Gyro suction gauge — the flight attitude indicator (5) and the direction indicator (2) are run by the gyros, which in turn need a vacuum to function. I can check if I have enough suction to operate the gyros

46   Nav and panel lights — I can turn on navigation lights at the ends of wings — port (left) is red and starboard (right) is green. I can also turn on lights behind my instruments

47   Emergency bus switch — for use in electrical emegency

48   Circuit breaker panel

49   HF High Frequency (long range) radio — for long range communication (extra for this flight)

50   Omega navigation system — (extra for this flight)

51   Omega track deviation indicator — (extra for this flight)

52   Selector switches for the extra instruments

1: SYDNEY TO CAIRNS

As I heard the engine roar into life and taxied my plane to the holding bay at Sydney's Bankstown airport, a great sense of relief swept over me. This was it. I was alone at last, crammed into the cockpit of my single-engined Piper Saratoga and about to begin my great adventure — my solo flight around the world. It was 3 August 1989 and the time was 8.55 am.

My last few hours on the ground had been hectic. Up at 4.00 am, a snack, then out to the airport to file my flight plan for Cairns and get my journey underway. I had prepared for so long that the actual flight had seemed like a dream for the future. Now it was really happening. And it was a perfect day for it, the skies clear and sunny.

My start-up checks complete, I radioed ground control. 'Golf Kilo Foxtrot taxiing at last IFR for Cairns.' I taxied to the run up bay, completed my pre-takeoff check: instruments, engine and controls. Everything OK. 'Bankstown tower Golf Kilo Foxtrot ready in the run-up bay.'

The reply crackled through my headphones, 'Golf Kilo Foxtrot track via the light aircraft lane. Golf Kilo Foxtrot cleared for Cairns.'

'Golf Kilo Foxtrot.'

The moment for which I had spent all that energy was

now before me. My children, Mimi and James, had come to the airport to see me off and close friends and well-wishers had crowded around me as I prepared to step into the plane. Media interviews had delayed my departure for an hour. I wondered if they realised what a sacrifice that hour was to me.

Now it didn't seem to matter. I was ready to go. I was here in my plane, actually sitting at the controls and about to take off. I taxied onto the runway, completed my line-up checks and pushed the throttle forward. My little plane responded. It was as if it knew that together we were about to experience an exhilerating journey. We rolled west on runway 29 and, as we gathered speed, we reached that part of an aircraft flight that is always a magic moment for me — the point of lift-off. It doesn't matter to me if I'm in a Jumbo or a light aircraft, as soon as I feel the wheels lift from the ground I experience an enormous pleasure, in affirmation of the wonders that we humans can create.

As my plane gained height and the wheels retracted, I banked right to pick up the light aircraft lane heading north to Hornsby and along the coast to Patonga, the first reporting point on my leg to Cairns.

I looked down over Sydney, laid out below me, and silently said goodbye to those familiar landmarks which I didn't expect to see for another six weeks. Media helicopters buzzed around me as television cameramen took footage to be shown on that evening's news. My radio crackled with the voices of several well-wishers who had flown in their own planes to escort me out of Sydney. Somewhere in the sky around me was my husband Neville's amphibian plane with Mimi and James aboard. Their voices came over the radio. It was a shock to hear them although I knew they were up there, accompanying me out of Sydney. It was one of those loving gestures that had drawn me to Neville in the first place, and, although we were separated, and in fact due for a divorce hearing while I was away, I felt very close to him.

As I approached Patonga and prepared to switch from

Visual to IFR flight rules which would take me inland, the voice of Kathy Salvair, a fellow member of the Australian Women Pilots, wished me well. My friends Lynora Brooke and Pat Wilson were on board with her.

I sighted Patonga and reported to Flight Service. I reluctantly said my goodbyes. It was obvious to me that Nev was having radio trouble as I couldn't receive his transmissions. I felt queasy as I particularly wanted to say a final fairwell to the kids, and hear their voices once more, but I had to let go of it and tuned in my navigation aids for West Maitland.

I checked my flight plan to make sure I was on time and to complete another of the hundreds of reports I would make during the course of the flight. I had prepared all the flight plans for the entire journey well ahead of time, although they often needed last-minute alteration. This one was, perhaps, the easiest, because I was flying over home territory, but it was one of the most exciting for me to prepare, and I have included it here.

At the top it contains my call sign, GKF (Golf Kilo Foxtrot), although outside Australia I would have to add VH (Victor Hotel) for Australia. It also notes the details of the type of aircraft, PA 32R or Z, followed by code letters for the type of radios and navigation aids on board. Next BK to CS is Bankstown to Cairns and the expected time of departure in Greenwich mean time or Universal time. The type of Flight Rules to be used on this section were Instrument Flight Rules (IFR), which simply means that I am qualified to fly in Instrument Meteorological Conditions, or cloud. The positions through which I am to fly are also included in an abbreviated form, for example, West Maitland WMD, Singleton SGT, Gunedah GDH, and so on. Where the asterisks are I made a position report to Flight Service.

The next column on the flight plan shows the lowest safe altitude to fly giving clearance above the ground of 1000 feet, the next column the altitude I intended to fly, then the true air speed, the magnetic track, wind direction and speed, magnetic heading corrected for drift, ground speed

CAA CIVIL AVIATION AUTHORITY AUSTRALIA

DOMESTIC FLIGHT PLAN
Original (White) - Pilot
Duplicate (Yellow) - ATS

GKF

125.8
125.7 Patonga
124.8

	AIRCRAFT IDENTIFICATION	CATEGORY	TYPE OF OPERATION	PRIORITY	PC	TYPE	Nr
FPL	GKF	(I) SP V	RPT CHTR AWK (PVT) MIL MLJ	SAR FFR MED VIP	A	P,H,3,2	

COMMUNICATIONS / NAVIGATION AIDS / SSR

| VHF (VR) VN | VHF HR HN | UHF UR | ILS 2 (L) LL | (F) | ADOF 2 FF | I VOR 2 O (OO) | DME X (D) | RADAR W | TACAN T | INS 2 I II | OMEGA 2 (M) MM | DOPPLER P | OTHER | NIL N | A (O) N |

DEPARTURE POINT ETD   FLIGHT NUMBER - (REPEAT AS REQUIRED) - FINAL LANDING POINT

13K (2200) CS   118.4 50 13   ATD: 2255   END OF DAYLIGHT   AT LOCATION

ROUTE SEGMENT/PFR	LSALT	FL or ALT	TAS	TR MAG	WIND	HDG MAG	GS	DIST	ETI	PLN EST	ATA	GS	EST next POS'N	Flt Proc, etc
X WMD	3100	3100	150	002	00/10	002	140	23	30	2325	31	45	124	80c+A
SGT	2600	A065	—	299	—	299	140	19	8	2343	43	2359		23
SCN	3400	—		313	—	313	—	38	16	59	59	0014		0cTank
Y GDI	5200	—	—	324	—	324	—	35	15	0024	16	31	127.1	
GDH	3500	—	—	327	/	327	—	36	15	0031	30	0049		
NBRI	3300	—		321	—	321	—	45	19	49	49			
X MOR	3700	—	—	350	/	350	—	48	20	0110	12	57	126.8	
SGE	2400	—	—	314	—	314	—	109	45	0167	51	37		
X RBY	2600	—	—	354	—	354	—	90	38	0233	36	585	R	7.20
EWL	5000	—	—	340	/	340	—	181	79	0355	52	17	122.3	
CMT	2300	—	—	3.9	—	319	—	57	24	0419	15	04		
OFT	2900	—	—	339	/	339	—	97	41	0500	56	0547	30	
X TL	5100	—	—	339	/	339	—	119	51	0547	34		30 TL	
TFL	6400	—	—	329	/	329	—	109	46	0620	14	24		
X CS	6400	—	—	332	/	332	—	45	19	0659	39			

FUEL CALC.	Min	L/Kg/...	Min		Min					CLIAS	Mach
Climb					57					110	
Cruise	466	125					11.04				
Atn											
SUB TOTAL	466	125									
Variable Reserve	60	09									
Fixed Reserve	44	10									
Holding (if req'd)	60	15									
Taxi											
Fuel Req'd	630	159									
Margin (cruise rate)		9									
ENDURANCE	630	168									
FROM	BK										

Operational Approval and Flight Plan Acceptance

SARTIME / NOSAR   DATE TIME   FOR ARRIVAL AT   TO ATS UNIT   BY (method)

ELB(A) L X   (LX)   SURVIVAL (FA) (LJ) (LR) (ER) (WA)   OTHER (Specify)   POB 1   ATS

PHONE NUMBERS AT LANDING POINTS   CAPTAIN *G Kennard*   COMPANY PVT.

DA 1641 (Rev 12/88)   Stock No. 16804.9

Gaby's first flight plan, from Sydney to Cairns.

according to the forecast wind, distance to next point, for example West Maitland, the estimated time interval to West Maitland. I would fill in the next column as I went along, with the actual time it took me, and then update my next position based on the last. This constantly changed as I moved through different air masses, but with each position report I was supposed to estimate my next time interval to within plus or minus two minutes.

On the lower section of the plan is the amount of fuel I had on board, with allowances made for a normal 45 minute reserve, plus 15 per cent of flight fuel, contingency for instrument flight and 60 minutes for holding in case of delays going into a primary airport such as Cairns.

I was on my way now and settling into the cramped conditions rather well, I thought. I looked about my cockpit with some pride. I thought it most important to organise the cockpit well, as space was extremely limited. To my right, in the co-pilot's seat, was auxiliary tank No 2. I used the top of it as a desk for my flight plan, maps and flight bag. It was mid-shoulder height and a bit awkward. I had my slide rule, calculator, pens, pencils, rubber, ruler, white-out, scissors and so on in a small purse inside the flight bag. Between auxiliary tank No 2 and my seat was a small space which contained the sound recorder and microphone for the Channel 9 cameras, which were mounted on the right front windscreen, one looking out over the nose and one looking at me, as well as a camera under the right wing and one on the tip of the rudder, looking over the aircraft. I had signed an exclusive agreement with Channel 9 and *60 Minutes* for my story, and they were to record the flight. I found the cameras very intrusive and often didn't turn them on. In this tiny space also there was the flap extension lever, the nose trim wheel and the emergency undercarriage extension lever. In what was left of the tiny space I had two litres of mineral water, some fruit and dry biscuits. I also stored basic tools such as a screw driver, pliers, rag, duck tape, etc in here.

Auxiliary tank No 3 and 4, directly behind me, were

about head height. On the top of these tanks sat my emergency equipment, my raft the closest thing to me. At the rear of the plane I stored my travel gear, but I was travelling very light with a couple of changes of clothes, underwear and toiletries only.

My flight to Cairns was uneventful although I discovered to my horror that one key part of my navigation system was not working properly. This was the Omega computer system which AWA Australia had kindly lent to me for use during this journey. I had especially flown my plane to Melbourne to have it fitted as I needed its accuracy to pinpoint my position at any time. I had been told that on long ocean crossings a four degree error in navigation would lead me to a watery ending, as I would be relying on picking up the ground navigation station alone. I tried to re-program the Omega, but couldn't do it. I was worried and hoped that AWA in Cairns would be able to help me.

I used this first leg to check my main and auxiliary fuel tanks and to make sure the selector valve that switched fuel from one tank to another was working correctly. I also calculated fuel consumption and engine performance, carefully assessing all the systems. This eight hour leg provided a perfect opportunity to do this. Along with the Omega malfunction, I discovered that there was an air bubble in my turn and bank indicator and that my directional gyro compass was out by 70 degrees. After all the time and money and worry I'd spent preparing the aircraft I was angry about the lack of attention to detail it had received.

It was a glorious day, blue sunny skies and the Australian countryside below was at its best. The rich farming plots, the rugged mountains and the tropical waters of the North Queensland coast combined to remind me that I wanted to return safely to this country which has always been my haven — my home.

As I approached Cairns the sky below me filled up with a layer of stratocumulus clouds which obscured my view of the countryside, but my initial radio contact with Cairns was greeted with enthusiasm and I made my first landing of the trip at 4.39 pm.

I had originally intended to take off the next morning on my first sea leg to Port Moresby, but I didn't want to leave Australia with any aircraft faults, so I decided that I would have to have my plane thoroughly checked over in Cairns, before I'd even left Australia. This was a bitter disappointment.

I taxied to AWA's service bay where I could hopefully get my Omega system operational, and then I would move it to the next hangar, that of a local company, Aircraft and Component Engineering, who would check my plane over and service it where necessary. Once I'd organised this and knew that I would have to delay my departure by at least twenty-four hours, I headed for my motel.

My hosts were Margaret and Norton Gill, both pilots. Nort was a local general aviation pilot and Margaret, like me, was a member of the Australian Women Pilot's Association. There is a real sense of camaraderie among pilots — male and female, light plane and big jet — and the eventual success of my venture was helped by the many people who went out of the way to meet me at airports, show me the sights of their cities, explain local flying conditions and weather patterns, and many other kindnesses, because we share in a fellowship of the skies. Margaret and Nort were the first of many whose willingness to help me played no small part in bringing me safely home.

Margaret and a group of local women pilots had arranged a tropical barbecue around the swimming pool later that night. It made me feel I was not too far from home. My friend Pat Wilson in Sydney phoned to tell me she was faxing my will to be signed the next day. My divorce was due to be heard while I was away and it made sense for me to have a new will. It was one of those last minute details that had escaped my notice in my rush to get away.

Although I was very tired from my long flight and the excitement of actually getting underway, I had trouble going to sleep that night. Many thoughts tumbled around in my head. I was worried about the Omega and remembered the trouble I had had when I flew to Melbourne by commercial airline to pick up my plane after the system was supposed to have been fitted. I had found two men arguing violently in the hangar about who should be doing what. There were wires hanging out of the plane and the work was far from complete. The men almost came to fisticuffs. In the end it hadn't been correctly fitted and on my return to Sydney I had had to have Australian Avionics check it over and do the best they could to get it right for me. Obviously, the initial faults were still there.

I also thought again about my reasons for undertaking the flight. After all, if I wanted to turn back, now was the time, as I had some perfect excuses at the ready. But I knew that I could never do that, because I wanted to make this trip too much, and I had put so much into the planning of it. I would never turn back.

It is simple to say, but difficult to explain properly — I had always wanted to fly. As a small child I was always fascinated by flying. I can remember when I was about six I would stand on a fence or wall and imagine that I could fly off into the wide world, just by flapping my arms. That dream never left me.

Amelia Earhart, who disappeared on a world flight in 1937, was my first childhood heroine. She was a famous aviatrix and the story of her disappearance while on a world flight was an intriguing mystery to a small girl growing up in the 1950s in Australia. I had first heard about her at school, but later I collected many books about her exploits.

And as I grew older and began to fly on commercial planes, both small and large, jets and propellor-driven, I simply loved the sensation. I loved the lift-off, the flight and the landings — all of it!

At the age of thirty-four, as a single mother, I decided the time had come to do something about this passion of mine — to start flying lessons. I scrimped and I saved, I gave up going to restaurants and theatres and movies and was able

to pay for enough lessons to achieve my Private licence, at
the South Coast Aero Club, New South Wales, in 1979.

In 1984, with the help of my second husband, Nev, I
achieved my commercial licence at the Royal Aero Club in
Bankstown, New South Wales. In 1985 I gained my Multi-
engine Command Instrument rating and in 1987 I received
my Seaplane licence. I hadn't done all of that for nothing!

But there had been more, much more to get me to this
stage. The planning for the actual flight had taken two years
and I had learnt such a lot. I realised for the first time that
you don't have to be good at everything, that you can ask
other people to help.

I expended a lot of time and energy trying to bring an
amphibious plane from the United States to Australia. Nev,
my husband, and I had bought the aircraft in America, but
he had not wanted to bring it home. He had had an accident
the year before in a seaplane, flipping it over in the Everglades
in Florida. He had had a lucky escape, not only from the
crash, but from the alligators, too. Understandably, he was
not anxious to try it again. In August 1987 I went to the
Lake manufacturers in New Hampshire to take delivery of
the new Lake Renegade Amphibian with the idea of flying
it home across the Pacific. Flying this aircraft across the
United States on my own was a big challenge for me with
my relatively limited experience. But when I finally arrived
in California, feeling rather pleased with myself and ready
to take on the Pacific alone, I was told that this aircraft was
not really safe to carry the enormous amount of fuel required
to fly it over the long distance from the United States
mainland to Hawaii.

It was a big disappointment to me, but as a consolation,
Jim Hazleton, well-known to many Australian flyers and very
much a pioneer aviator in his own way, offered to allow me
to fly with him back to Australia in the light twin he was
piicking up from the same place. I could then get a really
good idea of what long-distance ocean flying was all about.
And I certainly did. It was a wonderful and frightening
experience. I learnt so much more about navigating,

communicating and fuel management over long distances. Without this experience I doubt I would ever have had the courage to make my own flight.

It was during this flight that I began to think again about Amelia Earhart. My husband, Neville, had given me another book about her two years earlier, but I had really not taken in the fact that we were flying anywhere near where she disappeared until Jim said: 'It's about here that a lot of people think Amelia Earhart came down.' We were approaching our landing at Majuro on the Marshall Islands. From that moment on my interest in Amelia really developed and, on my return to Australia, I bought and borrowed every book about her that I could find.

Also, on my return to Sydney I determined to find a way to bring the amphibian home by shortening the distances between stops, thus decreasing the fuel load needed. Travelling around the North Pacific rim looked the most likely possibility, with stops in the United States, Aleutian Islands, USSR, Kurril Islands, Japan, Philippines, Indonesia, then Australia. I began to research the route and ask for permission to land in the various difficult areas, such as the Shemya American Airforce Base in the Aleutians, and the very sensitive Kamchatka Peninsula in Russia. I thought that with the softening of relations between East and West, Russia might well allow me to refuel on their territory, particularly as I was to arrive home at the time of our biggest air show, a Bicentenary event, in which the Russians were participating.

I received the all-clear from the Americans, but after many months the Russians refused. I was shattered. Almost a whole year of work was for nothing, or at least it felt like it. I really felt like a failure. Was I just a stubborn, persistent fool, not knowing when to give up?

Right at this time Nev decided to leave me for another woman, and my world seemed to collapse. I sank into depression and self-pity, really wallowing in my misery. But after a really vicious fight with Nev I pulled myself together and decided that I wouldn't let it get me down. I began to feel that I could take on the world.

Then began the second year of my preparations — preparations in earnest, for I had decided to fly around the world solo. This was a momentous decision, because I had no husband (who would look after the kids?), no aircraft, no spare cash, no job, nothing ... just a burning desire! I would show myself, and Nev, that I could do it, without financial help from him. I had a theory that if you could dream it, you could do it.

I made a strong commitment to my goal, and prepared an overall plan. I decided to raise funds for The Royal Flying Doctor Service, if I could, and that I would commemorate my childhood heroine, Amelia Earhart, by completing her world flight.

Still, looking at the whole project overwhelmed me to the point of paralysis. It seemed an almost impossible task. I had to break it into a series of steps. I took only one day at a time and tried to be as effective as possible for that day, knowing that eventually, if I adopted that attitude, I would get to departure day. I decided not to worry about the trip itself as that would be approached in the same manner, a series of flights A to B, B to C, C to D, etc. I had a strong inner belief there was something I had to do in my life, and this was part of it. One of my friends, John Baxter, said to me: 'You've got to do it. You can do it. If you don't give it a go you might as well give up.' I knew in my heart he was right. What other choice did I have?

But there was still so much to do. First I had to decide what sort of plane was most suitable, not only for my mission, but to suit my bank balance. After a great deal of research, I asked a friend in America to buy me a good, secondhand Piper Saratoga. It was a 1981 model, with a Lycoming 300 horsepower engine. It was perfect for the job. But it was American registered and had to be imported and converted so it could be Australian registered. I then had to have it extensively overhauled and prepared very specially for the project.

I had to plan my route in absolute detail and obtain permission to fly through all the different air spaces. I obtained landing rights and, where necessary visas, for the twenty-four countries I planned to visit. QANTAS were marvellous and helped with all the overflight permissions and landing rights, saving me thousands of dollars. I also alerted the Department of Foreign Affairs in case there were any problems.

I obtained topographical charts and instrument flight charts for all the countries and oceans I planned to fly over, and I read up on all the flight rules and procedures around the world. I prepared the flight plans in detail, with all the reporting points, altitudes, tracks and distances and proposed fuel consumption and so on. It took tens of hours of calculations.

One of the main considerations was the fuel consumption, not only ensuring I had enough money to pay for fuel along the way, but also to make sure it was available when and where I needed it, and that I had enough reserves. I had had a complex system of auxiliary fuel tanks fitted, to suit Australian standards.

I was also aware that the plane would need servicing along the way, and I would need to know where I could have this done reliably. I had to ensure, too, that I would have reliable weather information right around the world.

And I needed to organise things for myself as well. I needed to make sure that I was covered medically. I had to have inoculations organised and I packed special medicines in case of emergency. Like any tourist I had to try and organise suitable accommodation along the way. At home I needed to organise excellent child care so that I would always know that Mimi and James were well cared for and loved. I left them in the care of a wonderful woman, Katherine Lien, but I knew that Neville and Les (Mimi's Dad) and the neighbours would also do all they could and the kids would not be too lonely. I just had to overcome my feelings of guilt.

Finally, I had to work out how to pay for all of this. I borrowed money from the bank, using my house as security. I sold an old twin-engined plane, which needed a lot of work done on it, and I tried to raise money from sponsors. I was partially successful and I was so grateful for those people who had faith in me. Nevertheless, when I took off from Bankstown, I was more than $150,000 in debt. I couldn't turn back now.

But if I was going to make a success of my trip I'd need to get any rest that was offered, and, despite my overactive mind, the long flight from Sydney to Cairns finally took its toll, and I dozed fitfully until morning.

2: CAIRNS TO LAE

The next morning, I was feeling positive again, and determined to make a go of it. Nort drove me to the airport where I went to AWA about the Omega, to Clive Roberts about the instruments, and to Aircraft and Component Engineering about a general mechanical check. I can't speak highly enough of Colin Whyte and the crew at Aircraft and Component Engineering. They discovered a leaking propellor seal, which they replaced, and, most important for me, a split pin missing from the main strut of the left main wheel. Colin explained I was lucky to have landed without serious incident.

I felt the colour drain from my face. I remembered that Amelia Earhart, to whose memory I was dedicating this flight, had careened across the runway in Honolulu on the first leg of her journey, when the undercarriage gave way. I wondered if this was some kind of omen for me.

Fortunately, a new split pin fixed my problem. Amelia had had to have her plane shipped back to the American mainland and rebuilt before she could continue her flight. She also changed her original plans to fly east to west and her fateful last flight was in a west to easterly direction, the way I was going.

As these gloomy thoughts crossed my mind I reminded

myself that I hadn't crashed on my first landing, and thanks to the sharp eyes of Colin and his crew, I wasn't going to. I decided that if it was an omen, it was indeed a good one. This was borne out when Colin, who had worked all day on my plane, refused to take any payment for his work. 'Send me a photo', were the words this gallant Aussie man said.

This flight was already beginning to teach me many lessons, one of which was to learn to accept graciously the warm-hearted generosity shown to me by people like Colin.

That night I phoned my children in Sydney and, knowing that the 'bugs' on my first leg had been ironed out, I enjoyed a much more relaxing night's sleep. The next morning I was at the airport early to file my flight plan and clear customs. I was nervous. In a very short while I would be leaving the safe shores of my country and handling my first solo ocean crossing, 500 nautical miles over the Coral Sea to Port Moresby. Most single-engined planes take a more circuitous route that takes them over islands, thereby minimising the water crossing. I had had to receive permission from the Civil Aviation Authority to go the more north-easterly direct route over water, but I needed the practice, for soon I would be handling much longer ocean flights.

My departure went smoothly and as I headed out to sea my nervousness disappeared and I started singing. I was happy. My Omega was working and the yellow-green waters of the Great Barrier Reef reflected the sunlight back at me. It was a cheerful sight that lifted my spirits. There I was, singing away, checking my instruments and feeling pleased with myself, when I heard a loud whistling noise in the cabin. At first I thought I'd picked up a friendly ghost, but then I realised that air was getting into my cabin. I checked around and found the problem — the upper latch on the cabin door had not been correctly secured and was slightly ajar. The lower latch was the only thing holding the door shut.

'What a fool!' I thought to myself. This was another reminder to me that I could never be too careful in checking everything possible before take-off.

Securing the latch was going to present me with a

problem. The door that was ajar was the cabin door I had to use to get in and out of the plane. Getting in was a major exercise, for sitting in the co-pilot's seat was fuel auxiliary tank No 2, which I had to crawl over to reach my pilot's seat. So securing the latch was going to be very awkward from the inside.

I had to get out of my seat, lean over the fuel tank, and, with all my might, pull the door closed and secure the latch. I don't know how I did it, but I did. The effort took a lot of energy and I flopped back into my seat to catch my breath. I had surprised myself with the physical strength I had found to force the door closed, but in future I would make sure someone on the ground secured the door from the outside. It was a lesson well learned.

Needless to say, I did not resume my singing.

In an effort to regain my composure I turned to my instruments and carefully checked everything to make sure my craft was on course. This helped me to settle back into the flight. Then I began to notice how hot it was getting inside the cabin. Although I was flying at 7000 feet and the outside temperature was 15 degrees Celsius, the tropical sun was warming my cabin to the point of considerable discomfort. I knew I would have to make some provision for this as I neared the Equator.

One hundred and sixty nautical miles out of Port Moresby I became immersed in thick strataform cloud. 'Please don't let this be an instrument landing on my first touchdown on foreign soil,' I begged. The cloud persisted but then I descended out of it at 3000 feet to a visual approach. My prayer had been answered.

My landing was uneventful and I was hurriedly met by Air Nuguini's public relations manager, Geoff McLaughlin, who snapped some photographs of me and then took off. I taxied my plane to the local aero club and met some really wonderful people, including Chief Flying Instructor John Cleese, who had been there when my flying mentor, Ron Berry, from the South Coast Aero Club, had been a pilot in Papua New Guinea.

The people at the aero club were the high point of my overnight stay in Port Moresby. They were a group of eccentric aviators who had survived some of the toughest flying terrain and weather in the world and they reminded me of the barnstormers in the early days of aviation. They welcomed me with open arms and quickly had me seated with a cool beer while they explained the important facts I needed to know for successful flying in the tropics in general, and in Papua New Guinea in particular.

Invaluable to me were the instructions they gave on how to fly through the notorious Kokoda Gap, a part of my journey I had not been looking forward to for many reasons. The Kokoda Trail has many connotations for Australians who remember the Second World War, for it is here that Australian ground forces stopped the inexorable Japanese push towards mainland Australia. Also, the toll the Kokoda Gap has taken on fliers is horrendous and I had heard many of the stories of planes running into cloud or heading into a false gap and crashing into jungle-covered mountainsides.

I had to find the Gap because my plane was not fitted with oxygen and I could not simply fly over the Owen Stanley Ranges. My plane was equipped with the normal instruments, but I had borrowed a lot of extra gear for the flight, including the Omega 3100 Navigation system from AWA, along with the True Airspeed Computer for it. I had also borrowed a King 950 High Frequency long-range radio, and sophisticated survival equipment, a raft, flares, signalling devices, etc. As well, I was given a 'porta potty', which was a rather interesting lavatory device. What with the *60 Minutes* cameras and recording equipment and the auxiliary fuel tank systems needed for such a long flight, I simply did not have any more space, or weight allowance, for any more. I had made the difficult decision, therefore, not to take oxygen.

Thankfully, John Cleese drew me a rough map, something like the mud map a country person would draw if you asked them how to get somewhere. John's map might have looked rough, but it did the trick.

The other important thing I learned from these fliers,

was that I should take off early, particularly when planning to cross high terrain, and so avoid the heavy cloud build-up, which began mid-morning in these parts and was often followed by downpours and thunderstorms.

I determined to get out of Port Moresby early in the morning, and was in the air on my next leg to Lae by 7.30 am. My preparations had been orderly. I had filed my flight plan, which I had updated the night before in my hotel room, and re-checked the fuel, although I had refilled the night before. In the plane I had gone through what was to become my normal routine and almost second nature to me.

With the radio master switch off, I'd fire the engine up. For a cold start I'd do it in the following way. I'd open the throttle to a ½ inch, then put the master switch on, to start the electrical systems, then both magnetos, and then I'd put the electric fuel pump on, and prime the engine mixture to full rich, fuel pump on, then turn off the fuel pump and fully lean the mixture. I'd then engage the starter and move the fuel mixture to full rich. I adjusted the throttle to the correct setting and then checked the oil pressure. Then I turned the radio master on and the Omega on, because it took at least 10 minutes to warm up. My next job was programming the Omega, putting in all the way points to my destination. These are the positions on the earth's surface I expected to pass over. I'd put these in as longitude and latitude, and give each one a name, if it didn't already have one. This complete I was ready for my run-up and pre-take-off checks.

I would check the controls for proper movement, setting my directional gyro with my compass, setting the trims in normal take-off position. I would check that all the radio frequencies were correctly selected and at hand for quick future reference, both for the VHF and HF long range radio. I would ensure that all other navigational frequencies were correctly selected and all instruments operating as expected. I would then put my heading bug onto my expected outbound track, which simply means I put the pointer of my direction indicator onto my proposed outbound direction. Then I would run up the engine, check the magnetos, and exercise the propellor for full and normal operation, as well as checking the engine instruments to ensure they were reading normally.

Finally I would wait for taxi clearance from the control tower, airways clearance and take-off clearance.

Taking off from Port Moresby I had to climb to 11,000 feet to get through the Kokoda Gap and it took me some time, climbing higher and higher over increasingly precipitous terrain until I gained the required altitude. Generally I flew my plane at 8000 feet because it performed well at that altitude and oxygen is required for flights over 10,000 feet. I headed north-east for 42 nautical miles with John's mud map on my lap. My distance measuring equipment told me that now was the time to start looking for it. 'Yep, there's the dry, grassy lake, bank left then to the right.' And I saw the false gap. 'Yep, yep, this must be it. Bank here. This is it, the Kokoda Gap!' I saw a deep valley falling away beneath me and there, far in the distance, the sea. I had made it.

'Good work, Gaby!' I congratulated myself.

But just as I'd finished my self-congratulations I was brought up with a start. *Don't get too cocky, girl!*

Who was that?

Oh, don't worry. I just came along to keep you company and help make sure you get home.

Well, maybe it's stress and tension, I thought. I'll just ignore it. Perhaps it's only imagination. But it turned out to be the start of many conversations I had with myself on the flight.

I looked around at my plane. I had christened her the 'Spirit of Australia', and to me she was representative of many aspects of my country, from the gentle spirit of the Aboriginal people, the sheer strength and determination of the white pioneers who settled the land — and, importantly, the real stamina and courage of the pioneer women without whose support the men who 'tamed' the land could not have achieved their goals. She represented all this to me, and I was so proud of her. But, in lighter moments, I also nicknamed her 'Gertie', which was easier to say.

As I reached the coast and banked left on my north-west heading towards Lae, the cloud came rolling in and visibility was so poor I descended to 500 feet. I chose to fly about 500 feet off shore. I didn't want to suddenly find myself confronted with a jutting headland crowned by a jungle-covered mountain. Heavy tropical rain forced me down to 200 feet. I was practically skimming the waves, and I was scared.

I tuned to the Non-Directional Beacon (NDB) at Lae, hoping that my navigational skills were accurate and that my Automatic Direction Finder (ADF) would pick it up. I passed over a small island and studied my map. This must be Salamaua Island. If so, I was on course. I listened for the morse identification signal of the Lae NDB. It came on. I was delighted. Being on course was vitally important to me as I knew I would need all my navigation skills to find the NDB on my ocean crossing to Majuro in the Marshall Islands. My joy increased as the visibility improved and I flew into clear weather.

I banked left at Lae and headed into the Markham Valley, to Nadzab, the site of the new airport 12 kilometres up the valley from Lae. The Lae airstrip from which Amelia Earhart took off for the last time before she disappeared, and the one known to many United States and Australian airmen from the Second World War, is still there and now used exclusively by the military. Civilian traffic uses the new airport at Nadzab.

The tropical heat was fierce and, after a normal landing, I was met by Amanda and Doug Wright and their two children, Heather and Stuart. They did not know me, but had heard about my trip and driven out to the airport hoping they might see me. What a blessing that they did — they helped make my stay in Lae most memorable.

After lunch Amanda took me under her wing and showed me signs that Amelia's visit to Lae is not forgotten. The hotel I was staying in boasted an Amelia bar with photographs of Amelia and details of her last flight. Opposite the gates to the military airfield there was a memorial commemorating Amelia and her navigator, Fred Noonan, detailing their departure on 2 July 1937. The memorial had been overgrown but was cleaned up for my arrival.

Amanda also took me to meet Ella Birrel, a 75-year-old woman whose family had lived at Lae for three generations and who had therefore been in Lae when Amelia visited. Ella described to me the events that took place during Amelia Earhart's visit. It was especially poignant to hear this wonderful woman describe the last take-off, for Amelia Earhart and her navigator were never seen again.

Ella, Amanda and I made a wreath which we took to the memorial, intending to have our own small ceremony. It turned out to be a bigger event than I'd imagined. The memorial was opposite the gates to the military airfield and some personnel from the Royal Australian Air Force (RAAF) happened to be there. Our activities had roused their curiosity, so when they discovered what we were doing, they provided a small military escort. There we were, three women accompanied by a small contingent of military personnel from the Papua New Guinea Defence Force, and the RAAF. It was a very moving moment and I made a small speech about Amelia's exploits and her love of life, and we laid the wreath.

I was invited onto the base by the commander and was shown some of the old hangars still there. I was also driven up and down the runway. I could imagine Amelia, her Lockheed 10 Electra laden with fuel, her twin engines roaring to gain ground speed, waving out of the cockpit and smiling her engaging smile to the crowd gathered to wish her farewell, and lumbering down the runway to take off over the ocean and forever disappear from sight.

The next morning at Nadzab I made my pre-flight checks thoroughly. These were to become second nature to me, but they were most important, and I always did them, without fail. They involved looking at all external surfaces and moving parts, peeping in where possible to look for loose wires or nuts, looking for insects that might have found a home overnight. I also checked the antennae to see they hadn't come adrift, and checked the fuel for contamination. I checked the oil and carefully inspected the undercarriage and the tyres. I looked for oil leaks and loose spark plug leads. I checked engine nuts, the propellor and so on.

It all took me about 10 minutes, then, after shutting the rear door, I scrambled awkwardly over auxiliary fuel tank No 2 and into my seat.

I was granted permission to fly at 50 feet over the runway at the Lae military airport, giving me an opportunity to get a greater sense of what it must have been like for Amelia. As I flew down the runway I waggled my wings in a salute, a big lump in my throat.

My tears fell as I noticed a beautiful DC3 waiting for me in the skies over Lae, proudly showing its country's emblem — the spectacular Bird of Paradise. This wonderful plane escorted me for 100 nautical miles on my journey to Rabaul. I was sad when it peeled off to head back for Lae. Sad for many reasons.

3: LAE TO RABAUL

Today the skies were clear with occasional puffs of white cloud. The sea sparkled and the land below was a green carpet — almost idyllic. I was alone in the air again. I wondered if I'd ever get used to it. Most of the time I enjoyed my solo flight; it was wonderful — just me, my plane and the elements. Two hours out of Lae and heading towards Rabaul, however, was not one of those times.

The events at Lae had touched me deeply. I felt closer to Amelia Earhart and the person I understood her to be. I have read every book by her and about her that I could lay my hands on and the picture that I hold of her is of a tall, slim, elegant woman with an ethereal beauty and sense of style that seemed out of place in the times in which she lived. Yet, balancing this was her adventurous spirit and sense of fun. She lived in a time when men dominated, a time when being a woman was much more confining than now, when women had not long had the right to vote. Yet she firmly believed that women could do what men could do, and she did whatever she could to show that.

She was also a nurse, a very early social worker, a creative clothing designer and a talented poet. I had intended to name this book after one of her poems 'Courage is the Price', until I discovered that her sister had written a book about her by the same title. I have included her poem later because it very neatly sums up Amelia Earhart's philosophy.

But there was another reason why I felt low on my flight to Rabaul. The wonderful gesture by the Papua New Guinea Defence Force, in providing me with the escort by the DC3, brought up the itch of an old emotional scar with which I thought I had dealt.

I was born at the Mercy Hospital in South Melbourne, Victoria, in 1944. My birth certificate shows my mother as Patricia Marguerite Buchan, and my father as Thomas Dudley Dwyer. But when I was eighteen, my mother, during a terrible tirade, cruelly let out that Thomas Dudley was not my father.

I'll never forget that day. I was home from nursing for a few days and Sheri, my younger sister, and I were sitting on a sofa in the loungeroom, cowering from our mother, who was in a rage and directing her invective at me. Sheri was shocked. I wasn't. In many ways it was a relief to know and understand, because I had never felt a part of Thomas Dudley, though I loved him very much.

My mother and Thomas Dudley were married six years before my birth when Thomas, as a young man, volunteered for the Army. It was a time of world war, and the chances of him returning seemed remote. My mother sat out the war for a year, and then, flying in the face of the social conventions of that time, took herself to Brisbane and joined the staff of the American forces stationed there.

The inevitable happened. He was a handsome young American pilot with a wife and child back home. They fell in love, but from what my mother told me it was not the usual storybook romance. She loved him as she had never loved anyone in her life.

Then two things happened which changed the lives of all involved completely. The Army notified my mother that Thomas was coming home. He was shell-shocked and probably suffering from what today is recognised as the trauma that sensitive people suffer when they are forced to continually deal in slaughter.

The other occurrence was me. My mother was pregnant.

She was faced with a choice which was to affect not only her life, but many other lives. On the one hand she was loved and wanted by the man she loved, her American flyer, my father. On the other hand she cared for and was needed by Thomas Dwyer who, traumatised by the war, had clung to the thought of returning to her as the one thing to keep him going.

She chose Thomas. My father pleaded with her to stay with him, but she was resolute and returned to Melbourne to be with Thomas. I don't know if she told him about my conception. If she didn't at that time, I suspect that later she did.

Not long afterwards my natural father was killed. The plane in which he was flying crashed. It was a Dakota, the military version of the DC3. Seeing that wonderful plane that had been so diligently cared for by the Papua New Guinea Defence Force made me feel melancholy and frustrated — melancholy because I had never met the man who was my father, and frustrated because I had not even known of his existence until I was eighteen.

I had felt very close to my father when I had begun my flying instruction, and in an odd way I knew he'd be proud of me, that I was doing something I really loved. I felt I was earning his approval. Now, here I was flying over the jungles of Papua New Guinea, jungle very similar to the one in which his plane had crashed, with him in my mind. I saluted him.

But I knew I could not afford to be maudlin. My flight demanded all my attention. It was only a short three hour trip to Rabaul, which I had decided would be my jumping off point to Majuro in the Marshall Islands. I could have gone direct from Lae, but that would have meant a 14 hour flight into darkness, and I wanted everything possible going for me. I was very concerned about finding Majuro, a coral atoll 30 kilometres long in the vast Pacific. I decided that departing from Rabaul would save precious time — and my nerves — on this, my first major ocean crossing. In many ways I wish I had not made that decision.

I had planned to stay overnight at Rabaul and get a very early start the next morning. A friend of Doug and Amanda Wright's, who operated a diving school in Rabaul, came to meet me and I stayed the night at the house he shared with his two associates.

At 4.00 am the next morning I arrived back at the airport, went to the office and filed the flight plan for my leg to Majuro. I walked to my plane and got in, preparing to settle myself ready for the flight. I reached behind me for the carry-all which held my emergency equipment, intending to put on my life jacket, something I did before any water crossing.

It was not there. I couldn't believe it. I got out of the plane and walked around and unlocked the back door. When I opened it, I saw immediately that my plane had been broken into and ransacked. I went though all my gear and found that the thief, or thieves, had also taken other valuable items but, of most importance, they had taken the one thing I needed for this major water crossing. I could not possibly leave without basic survival gear. My own instinct told me that to do so would be foolhardy, and I had no intention of being without it. I also felt that if I had the equipment I wouldn't need it. My emotions were a mixture of tears of frustration, disappointment and shock. What a setback!

The airport security people were of no help and showed hardly any concern. I wondered what it was they were employed to keep secure.

After regaining some of my equilibrium I took stock of my situation. I still had my inflatable raft, but all my other survival gear was missing. I had been lent a special life jacket which came equipped with everything I would need if I ditched in the sea. It had special pockets and equipment attached which included flares, lights, mirrors, waterproof matches, flashing strobes and a survival beacon which would give off a special locating signal. I felt particularly bad about its loss because it did not belong to me.

Also missing were a pair of field glasses, the spare vacuum pump, spark plugs and the plug wrench. I imagine that now most of this gear is sitting on the altar of some cargo-cultist in the jungle around Rabaul.

I cancelled my flight plan and phoned Sherry Stumm in Sydney. Sherry had tried to get sponsorship before I left Sydney and now she was helping with publicity. She boosted my flagging spirits and I knew that this was one more test that I would have to overcome, but must admit that this was one time I felt like giving up.

I went to the office of TAL Air, the local airline and told my tale of woe to the people there. When I was in the office, I noticed the office safe with its door crudely blown off. 'We keep it like that', the manager said. Apparently it had been blown open twenty-five times.

The TAL Air people were very sympathetic and said they would scout around and see if they could find replacement gear for me. The people at Pacific Helicopters also said they would see what they could find. All I needed were the minimum requirements of the Australian regulations — a vest, a survival beacon and a raft, which I already had. I wondered how the thieves had missed it.

I booked myself into a hotel close to the airport, for even if the replacement gear were found that day, it was too late to head for Najuro. I needed to leave at first light to make the 1400 nautical miles to Majuro in daylight.

Thanks to the efforts of the staffs of TAL Air and Pacific Helicopters, I had my replacement survival gear by three that afternoon. It wasn't as technically advanced as the gear that was stolen, but it met the requirements of the regulations. I planned for an early departure the next morning and made a point of asking for a 4.00 am wake-up call. The emotional strain of the day caught up with me and when my head hit the pillow that night I think I was already asleep.

My eyes popped open and, though it was pitch dark, I knew something was amiss. I looked at the clock. It was 4.50 am. Rabaul had done it again. I was dressed and ready to leave in 15 minutes. Not a soul stirred as I ran down the half-lit corridors searching for the night clerk. I finally found him and he glibly told me that he had received no note to wake me up. I asked him to drive me to the airport and, after what seemed like a long search for the keys, we drove off. I promised myself that one day I would laugh at it all.

We arrived at the airport. It was 5.30 am. I hurriedly filed my flight plan and, getting the weather report, ran to my plane, thoroughly checking it for signs of further theft. I put on my life jacket and attached my raft to my wrist with a longish lead. Tony Tobler, a very helpful officer in the Air Force, had once said to me: 'The only thing you end up taking with you if you have to ditch in the sea and manage to get out is what is attached to you.' I never forgot those words.

I completed a thorough preflight check and at 6.09 am, only 39 minutes late, my wheels left the ground and I thankfully said goodbye to Rabaul.

Although the events of the last day had probably soured me a little, I've got to say that Papua New Guinea sadly disappointed me. I remember hearing stories of how beautiful it was, in particular Rabaul. Although the natural beauty was there, it had obviously changed.

4: RABAUL TO MAJURO

My estimated time en route to Majuro was 11 hours and, if all went well, I would make it before dark. Flying east over such a distance meant I would lose about 2 hours of daylight. Once in the air I set my course for Buka Island, north of Bougainville — the last land until Majuro. With Buka behind, my head — drawn like a magnet — kept turning back to land.

It's okay Gaby, you can let go of the land now.

So began the advice I was to give myself throughout the journey. I turned my head and looked forward to the vast reaches of the Pacific.

I received a call on my HF (high frequency) radio. Thanks to the Australian Overseas Telecommunications Commission, I could make and receive phone calls, atmospheric conditions permitting. I cried to hear the sound of a human voice. The journalist, who was filling in for the ones I normally spoke to had no idea of where I was, or how I was affected by my first major water crossing. I guessed he was safe in a city studio with people buzzing around. He cheerfully signed off, leaving me crying my eyes out.

I got out my pictures of the kids and had a good look at them. I then pulled myself together, checked my calculations, wriggled my toes and stretched my arms out as far as I could. I was going to be all right.

The tropical sun was very hot and I covered my face with sunscreen to block it out. Although at my cruising altitude of 8000 feet the outside temperature was about 22 degrees Celsius, the temperature inside my cabin was over 40 degrees Celsius. How I wished for the air-conditioned luxury of a more sophisticated plane.

Mary O'Brien, a friend in Sydney, had lent me a sun shield which I placed on the inside of the windscreen. That piece of black cloth with its silver backing was a blessing and even if I couldn't see outside, it was cooler with no glare. I could see my instruments, which I was using to fly, but the situation seemed ludicrous to me. Here I was, with daylight all around me, not looking where I was going. I started to laugh. Mary's wonder cloth had helped to dispel the gloom I had previously felt, and my spirits picked up.

The hotel in Rabaul had packed me some sandwiches which were wrapped in foil, and I thought I would have some breakfast. My laughter increased when I discovered that the cabin temperature was so hot that the sandwiches had baked and, regrettably, were inedible. So I settled for hot banana and drank some water.

I flew through the equator and had my first taste of northern latitudes, which really didn't seem all that different from southern ones, certainly not in the equatorial regions. I settled into my flying, checking my instruments and, most important, peeking through the sunscreen to study the sea.

I had learned on my previous flight across the Pacific with Jim Hazleton, that the ocean says a lot. The chop on the waves indicate the direction and strength of the wind, while, if it is smooth, then the wind was dead calm or slight.

The heat in the cabin, combined with my apprehension about finding the Majuro NDB, meant that I was literally sweating — believe me, it was too hot and I was too nervous to perspire delicately. I kept my eyes glued to the Automatic Direction Finder (ADF) and, with my aural identification button switched on and the volume turned up, I listened intently for the morse signal.

At 100 nautical miles out the needle flickered and then pointed dead ahead. I was overjoyed and heard the morse signal as I realised that in only a short time, my first long ocean crossing of almost 1400 nautical miles would be completed. About 50 miles out of Majuro, the voice of Majuro tower came through my headphones. I identified myself and prepared for my landing.

I passed over Kili and Mili Atolls and noticed what looked like a small airstrip near the beach on one of them. Many people believe that Amelia landed near Kili Atoll and I wondered if it was possible to land on that strip.

As I approached Majuro the light was dimming with the rapidity that announces night in the tropics and I passed through towering cumulus cloud formations. I hoped I wouldn't have to make an instrument approach. My luck held and, as I got closer, the voices of two American pilots came through my VHF radio. One of them, a woman who was obviously flying out of Majuro, was telling the other, a man who was landing ahead of me, that she was pleased to be leaving 'that cockroach ridden hole'.

'Cockroaches or not,' I thought, 'if that lady only knew how pleased I was to see it!' The clouds broke up during my descent and, at 2000 feet, I saw the airstrip a mile ahead in front of me. I congratulated myself and the lecturers who had drilled me in navigation. I had found this small piece of land about 30 kilometres long and 320 metres wide out of the vast ocean.

The wheels touched down. *Good on you, Gab. Good girl. You've done it.* It was 6.45 pm local time and the sense of relief was tremendous.

The light was fading fast and I taxied to the parking bay and killed the engine. Joe de Brun, the airport manager, came to greet me. However, not even his cheery greeting could ease the pain of paying my landing fees, customs duties, control fees, and the like, which cost me US$500.00. I don't think I'd like to pay for landing a Lear jet on the Marshall Islands.

I had the plane filled up with Avgas and then the fuel man gave me a lift into town. I found the main hotel booked out and was told that my only chance of a bed there was to wait and see whether a 9.00 pm booking showed up or not. I was very tired and had not properly eaten since dinner the night before. I booked myself into the Gateway Hotel down the road. It boasted 'bungalow-style' accommodation and, when I reached my hut, I found evidence of the previous guest. I'm normally very fussy about things like that, but that night I couldn't have cared less.

I freshened up and walked to the restaurant, where I ordered a nice cool beer and a meal. The condition of my bedroom made me apprehensive about the meal, but I needn't have worried. It was very enjoyable. A tropical downpour was thundering away outside and, with my feet on terra firma and a good meal, I felt happy. I ordered another beer and was joined by a young Marshalese girl, about seventeen. I enjoyed telling her about my family. But I couldn't keep my eyes open any longer, so I excused myself and went to bed. Even the lumpy mattress looked good. I don't think a hurricane would have woken me up that night.

I was up and showered by 8.00 am and decided to forego breakfast and walk to the Airline of the Marshall Islands office, to find out about the weather reports and how to file the flight plans for my next leg to Honolulu. My logic told me it must be on the one, long main street. My logic didn't tell me how far along the main street it would be. By the time I found the office it was very hot and I felt faint. I think the lack of breakfast, the high temperature and humidity, combined with my exhaustion from the day before, were taking their toll, a sign that I should take it easy.

As I leaned on the wall of the office, I spotted a very busy man. I got myself together and approached him. 'Excuse me, would you be the manager?' I asked.

Something in my appearance must have struck a responsive chord. He stopped what he was doing and

introduced himself. Neville Hill was his name and meeting
him at that moment was a godsend. He is an expatriate New
Zealander and my time with him was very fruitful, in many
ways.

When I told him the story of my flight and the
connection with Amelia Earhart, he arranged for me to meet
the Chief Secretary of the Marshall Islands later that day. This
was a singular honour as the Chief Secretary is the head of
government and, in many ways, like royalty, and to get to
see him was no small achievement. I didn't realise that
meeting him was to interest me further in the mystery
surrounding Amelia's disappearance.

Neville took me to lunch and then drove me to the
administrative buildings where I met the Chief Secretary,
Leonard de Brun. He told me that when he was a young
man of seventeen, during the Japanese occupation, his father
had come home from work and told the family that he had
seen an American female pilot with short, blonde hair, and
that the Japanese had taken her away.

This helped to confirm for me the stories that Amelia
had landed, out of fuel, near Kili Atoll, one of the Marshall
Islands, and, for reasons unknown to me, the Japanese had
considered her a spy. (Before the War, the islands were
mandated to the Japanese, who occupied them and then kept
everyone else out.) I believe that she eventually died,
probably from dysentry, or similar, on the island of Saipan,
which was the regional military headquarters, and that her
navigator, Fred Noonan, was probably beheaded. According
to the books I have read, the ships that the Japanese said
were searching the area for her were still in harbour in Tokyo.

Further confirmation came later that afternoon, when
80-year-old Oscar de Brun came to see me at Neville's office.
Oscar, who was Leonard de Brun's uncle, told me that when
he was twenty-eight he had definitely seen Amelia Earhart.
To have an eye-witness account given by a frail old man who
had gone out of his way to seek me out and tell me, was
certain corroboration as far as I was concerned.

Neville Hill's generosity was extended further that

evening when he took me to his home to have dinner with his wife and two beautiful children. It was great to have a meal with a family and I couldn't help thinking about my own two children all the way back in Sydney. Neville took me to my hotel and we arranged for him to pick me up at 4.00 am the next day so I could get an early start on my long sea flight to Honolulu.

As I dropped off to sleep that night, I thought, cockroaches or not, I sure am pleased I came to Majuro.

Neville was a little late picking me up the next morning. It was very black and we were in the middle of a tropical rainstorm. I filed my flight plan, doing my wind calculations by torchlight. I couldn't get an accurate weather report and my chief concern was having enough fuel to make this long hop. There was a ferry pilot at the airport who was preparing to fly west to Guam. He looked at the direction I would soon be going and told me he was glad he was flying west. I knew what he meant. The sky to the west was clear. To the east it was grey, with flashes of lightning.

As was normal for this particular leg, I was well over the maximum gross take-off weight, but I thought I managed this overloaded take-off well, although I was soon climbing through heavy cloud and being buffeted by strong winds. Within 100 nautical miles I had flown into clear blue skies and settled down for the 1900 nautical mile long haul to Honolulu.

5: MAJURO TO HONOLULU

My concern on this leg of my journey was getting the best possible performance out of my plane. In Australia, as in most developed countries, acquiring fuel and fuel management is not a critical factor for light aircraft. It is always possible to plan a journey, no matter the distance, to allow for refuelling stops along the way, to get out and stretch your legs, and possibly take a meal break.

My flight across the Pacific allowed me no such luxury and I had to get maximum efficiency out of my plane to ensure that I had sufficient reserves of fuel necessary, and a margin of safety.

I will explain how my plane was laid out so you can get some understanding of the range and performance. First, I need to say that all my measurements are calculated in nautical miles and United States gallons. Although most of the world now uses metric measurements on the ground, the fellowship of the air still works in the old system. A nautical mile is the equivalent of 6080 feet and a statute, or ground, mile is 5080 feet. So a nautical mile is the equivalent of 1.15 statute miles or 1.85 kilometres. A United States gallon is roughly equivalent to four litres.

As I have already partially explained, I had a number of auxiliary fuel tanks fitted in the plane, and the diagram

I have enclosed will, I hope, explain how these were laid out.

The fuel I carried in them was configured in the following way:

Left wing (main tank)	50 gallons	3 hours flying
Right wing (main) tank	50 gallons	3 hours flying
No 1 Auxiliary (in nose)	27 gallons	1.5 hours flying
No 2 Auxiliary (beside me)	38 gallons	2.5 hours flying
No 3 Auxiliary (behind me)	48 gallons	3 hours flying
No 4 Auxiliary (behind me)	48 gallons	3 hours flying

Of course, the hours figured are very approximate and relate to ideal conditions only.

The auxiliary tanks were vented, which meant that they were fitted with hoses at the top which fed into a hose on each side leading to the outside of the plane. This ensured that no fuel vapour ended up in the cabin and that the pressure in the tanks as the fuel was used enabled the fuel to feed to the engine properly. There was a small V-shaped crawl space between tanks 2 and 3 which, by crawling over valves and pipes, allowed me access to the rear of the plane where I had a small space for luggage. There were several fuel drain valves under the fuselage which allowed me to check fuel for water and other contaminants.

To meet with Australian Civil Aviation Authority approval I had my auxiliary tanks designed and constructed to meet with certain safety requirements, the main concern being weight and balance. In my aircraft each auxiliary tank fed directly into the engine, as can be seen in the diagram. This differed markedly from the much simpler system used in the United States where the auxiliary tanks feed into the main tanks in the wings and then into the engine.

I chose the Australian system because, as an Australian woman, I wanted to fly an Australian registered plane. It would have been about $20,000 less expensive if I had chosen to fly using the American system, and, as events turned out, less frightening.

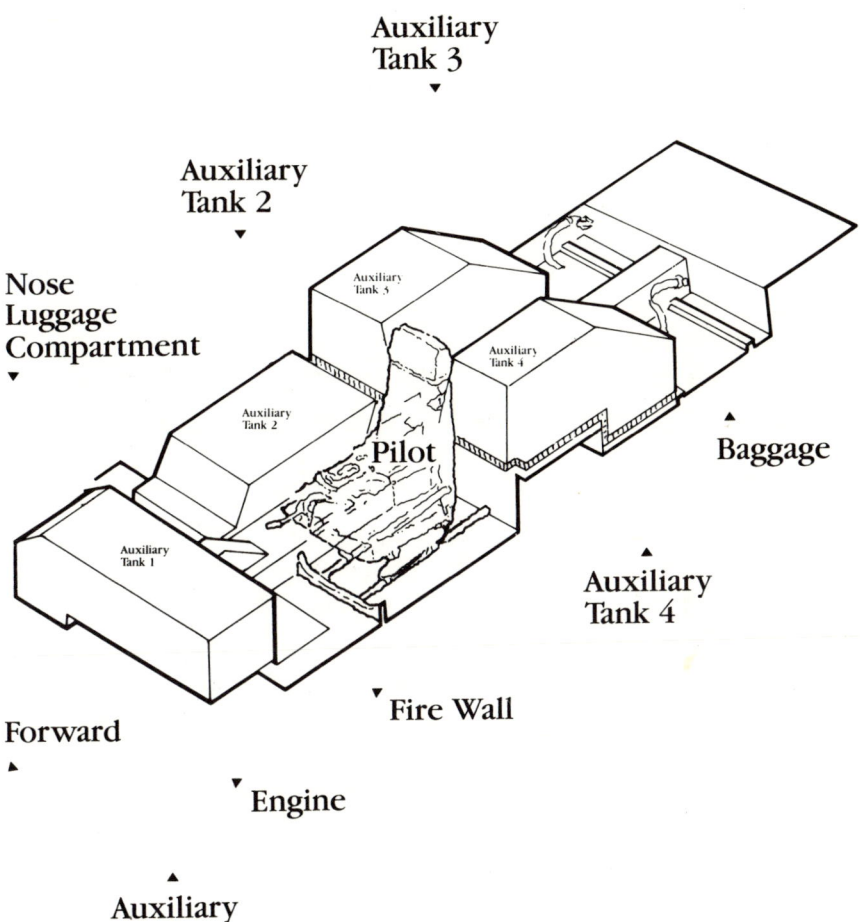

Auxiliary Fuel Tank Arrangement

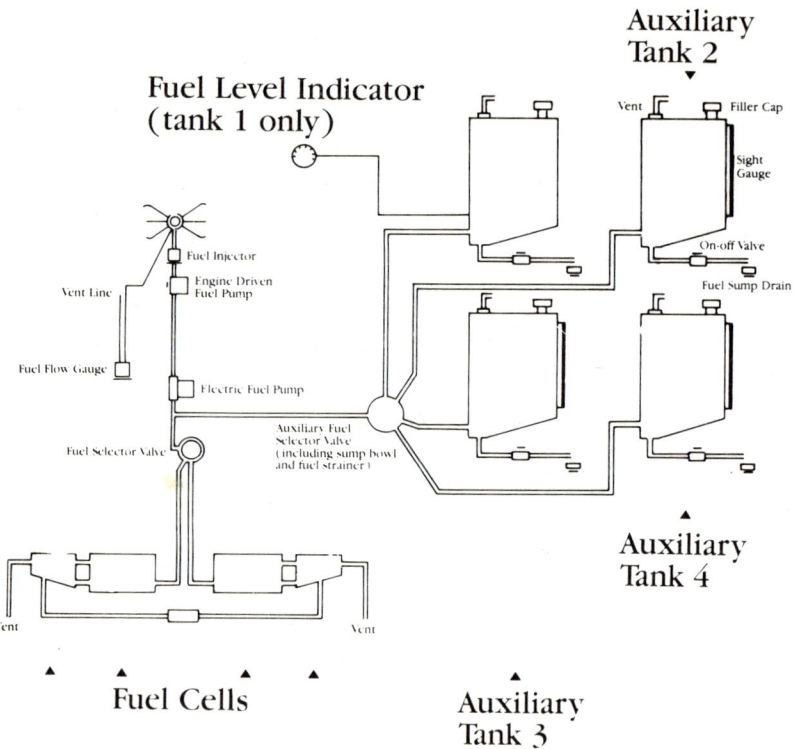

Long Range Fuel System Schematic

A plane's efficiency is dependent on many factors, some to do with the physical characteristics of the plane, and some, like wind strength and direction, outside the pilot's control. The cruising altitude, the amount of power being used by the engine and the fuel/air mixture feeding into the engine, along with the plane's trim and centre of gravity, were some of the main factors I had to take into consideration on the long legs where it was essential for me to gain the maximum performance in order to reach my destination with enough fuel in reserve for contingencies.

The trim of the plane is similar to the position in which the skipper of a sailboat puts the boat to gain the maximum advantage of the air flow across the sails. The trim of a plane is vital. If the position alters so that the nose is slightly up and the tail down, more drag is created and the plane is said to be at an inefficient 'angle of attack'. So it was important for me to use my fuel in a way that would keep the plane's centre of gravity and trim within acceptable limits. Before I left Australia I was told that it is best to use the fuel from front to back. I had to work out the best way to do this in my plane.

Before leaving Majuro I had calculated that, considering the wind, by using 65 per cent power and flying at 8000 feet I would have 14½ hours flying, actually 14 hours and 37 minutes. My plan was to fly over Johnston Island, an American Military Installation 700 nautical miles west of Honolulu, and use it as a navigation check. It is literally a pin-prick in the vastness of the Pacific Ocean, and I wanted to be sure of my position.

My planned flight interval to Johnston Island was 9 hours 20 minutes. I spent most of the flight checking the instruments, checking the fuel mixture and the trim of my plane and calculating fuel consumption. I also spent a lot of time watching the ocean and deducing the strengths and direction of the wind from the waves. I was flying into very strong head winds now, which I estimated were gusting up to 60 knots.

This was heavy duty, and I was worried.

The 9 hours and 20 minutes I had estimated before sighting Johnston Island came and went and the horizon showed me nothing but waves. Another 30 minutes ... 45 minutes ... an hour went by and there was still no sight of it. I became really worried, wondering if my navigation was out, and concerned about my fuel. Would I run out of fuel and end up ditching in the sea?

'Surely', I thought, 'I should be picking up the NDB on the island by now — unless I'm too far off course.'

I decided to contact Honolulu by radio. I announced my international call sign: 'Honolulu this is Victor Hotel Golf Kilo Foxtrot.' The return call came in and I asked if the NDB at Johnston Island was operating that day. It was a Saturday and I thought the military base might have been operating with a smaller staff and forgotten to turn it on — or at least I hoped so!

I don't know if my surmise was correct, but at exactly one hour and 15 minutes beyond my planned time, the needle of my ADF swung straight on the nose. I was right on course. Johnston Island was 28 nautical miles ahead of me.

I breathed a sigh of relief and then began to check my calculations. I was worried that the extra time taken fighting the head winds had used up my fuel reserves. My calculations showed that I had just enough fuel to make it to Honolulu, which was a further 5 hours and 17 minutes ahead, and if the headwinds continued I could add another 45 minutes to that. I had enough to get there with no reserves and no allowance for headwinds.

I had a problem.

It was then that I made a decision to land on the island. I couldn't risk ditching in the sea just short of the Hawaiian Islands, although landing on Johnston Island presented a problem. It is a high security military installation and it is not permissible to land there without clearance, unless an emergency situation has developed.

I was in an emergency situation.

I tried to raise the island tower, but received no

response. *Okay, Gaby, let's call Honolulu and tell them.* Yes, let's, but let's do it as professionally as possible. We don't want to be embarrassed about all of this.

'Honolulu, this is Victor Hotel Golf Kilo Foxtrot. I have a problem. I am worried about my endurance. In fact I am very concerned and I have decided to land at Johnston Island.'

The reply came back: 'Does that mean that you are declaring an emergency?'

'Yes, I am declaring a fuel emergency! Can you let them know that I am coming in.'

The reply was droll. 'I'm sorry, you'll have to contact them yourself.' She gave me the Johnston Island tower frequency.

By this time I was circling the landing strip and another 15 minutes of fuel had been used up. I had no response from the island. It was late afternoon and I had been lucky to have spotted the atoll through a break in the clouds. I was almost into my final approach, so I contacted Honolulu again and told them that I couldn't raise the island. Their reply: 'Keep trying.'

It was then that I saw some people race into the tower. By now I was on my final approach. I called them. 'I'm very sorry to have to do this, but I've got a fuel problem.' I think I got the words out only a few seconds before my wheels touched down.

A military vehicle with a large 'FOLLOW ME' sign suddenly appeared at a taxi apron, and I obligingly followed as it headed towards a group of buildings. I could see about eight armed soldiers rapidly taking up positions behind barricades around the edge of the parking bay. They were pointing their rifles at me! *My God, you've done it this time, girl.*

I thought of Amelia Earhart and wondered if I would languish away, hidden from the world in a military jail somewhere. 'Is this real?' I kept asking myself. It certainly was.

I shut the engine down and a very stern and overweight

officer of some sort approached me. He was not smiling. I really think he had forgotten how. His surly countenance increased my apprehension. *Oh gosh! We're in for it now, Gaby!*

I timidly opened the storm window. A gruff voice ordered: 'Disembark and stand by your aircraft.' I noticed the absence of any pleasantness. There was no 'please', or 'kindly'. I don't think I would be made for the military.

So, there I was, crawling over my auxiliary tank — hot, sweaty and feeling just plain 'yuk', and all this mixed with several degrees of terror. I finally got the door open and disembarked. It was not one of my most famous disembarkations. No pause with the door open to smile and wave to the crowd. There was nothing heroic about me as I stood by my plane, sneaking glances out of the corners of my eyes at the guys behind the barricades, who were popping up with their rifles pointed at me, then popping down again, playing some sort of game of musical barricades. This is no game, I reminded myself.

The surly officer barked: 'I would like to search the aircraft'. Later, I wondered what would have happened if I had said no. Fortunately, I didn't. Only long distance has given me the luxury of formulating clever answers to his rudeness. Right then I was on automatic.

On cue, with the surly guy finishing the inspection of my plane and finding no hidden devices of a subversive nature on board — although I did wonder what he thought of my porta-potty — the camp commanding officer came up, in a pair of jogging shorts. What a contrast. He was a pleasant man, who called me 'ma'am' and treated me with dignity.

I was escorted into a waiting room, flanked by the commanding officer, the surly officer and two women personnel, one of whom was black and who, as I caught her eye, gave me the most fleeting of grins. If that woman only knew what that grin did for me at that moment.

I continued talking to myself. *These people are Americans — our allies. They wouldn't want to hurt you*

at all. Besides, you don't think they'd risk an international
incident over someone as lowly as you, do you?

The interrogation from the officers slowly changed to
questioning and then drifted into conversation. The
commanding officer asked me if I'd like an ice cream. They
gave me two, and two cans of lemonade.

The airfield was set up to service jet planes and so
carried jet fuel. They searched around and, thank God, they
found a 55 gallon drum of Avgas, which they hurriedly
transferred into my plane. Once they had established my bona
fides, they wanted me out of there as quickly as possible.
I was an embarrassment to them, and outside the normal
operations of the base. But I was allowed to phone Honolulu
first and tell the people who were expecting me there that
I was delayed.

It was coming onto night, and they told me they had
no night facilities on the strip. So I was very quickly escorted
back to my plane and, as I climbed to my cruising altitude,
the last pink glow of sunset slipped away and it was dark.

I was ecstatic. I had enough fuel. I trusted my navigation,
and the winds had dropped. I contacted Honolulu centre
on my radio. Looking back in the relative security of my plane,
I could see how funny the incident on Johnston Island really
was.

My amusement continued when I used the porta potty.
For some reason I had never managed it well, but this time
I made even more of a mess of it than usual. From then
on I called it 'the beast'. But at least I didn't get myself
completely wet as I had on my first leg, from Sydney to Cairns,
when I had had to use my jacket to wrap around my waist
before I could get out of the plane. I never did master 'the
beast' until the flight from Darwin to Alice Springs on the
way home — a bit late!

I established contact with Honolulu radio and spoke to
them on my HF radio until I was within range of Honolulu
Flight Service, whom I contacted on my VHF radio. Communicating with the VHF or line-of-sight radio is always a
lot easier and, once I had made the initial contact with

Honolulu Flight Service, my mind was eased quite considerably. This usually happens about 50 miles out and it meant that I was pretty close to my destination. My heart always felt a lot lighter then, particularly at night over vast oceans.

Exactly five and a half hours after leaving the tiny atoll, I saw the lights of Honolulu sparkling in the night sky. After all that ocean it was like a fairyland sparkling for me.

I was transferred from Flight Service to Honolulu approach control frequency, who then, a little later, transferred me to Honolulu tower. I was radar vected onto final approach and directed to make a visual landing. Tower asked me if I had the runway in sight. For an instant after they asked me I couldn't see it, but, after what seemed like hours, but was only seconds, I spotted the high intensity runway lights. I continued my approach and made a perfect landing.

I touched down on Honolulu at 2.00 am local time — exactly 16.5 hours after my take-off at Majuro.

The international airport at Honolulu is big, and in the middle of a black night is not the easiest place to find your way around, particularly if you're not used to it. I asked Ground Control for directions to a parking bay and was told, for my trouble: 'Ma'am, our job is to get you on the ground.' Perhaps he was a cousin of the surly officer at Johnston Island. I badgered until I got some hazy directions and then, after taxiing over grassy strips, I finally found the general aviation area. It was pitch black. A news crew from Channel 9 in Sydney was supposed to meet me. I thought they had probably given up waiting and gone to bed, which I thought was a good idea. I contemplated bedding down in the plane until daylight, when I would be able to see to get myself out.

Just then I saw some car headlights in the distance. They drew closer. It was a news crew. I told them of my Johnston Island experience, and then they asked me to do some taxiing, so they could get some recorded on tape for future use. A car from the United States Customs Service then drove up and, when the news crew had finished, I was driven to the Customs area at the International Terminal and processed. Thankfully, no large flights had landed and I got through customs very quickly. I didn't have much to inspect and nothing to declare.

Channel 9 had booked a room for me at the Sheraton Hotel. As I opened the door I saw a most inviting sight — the bed. It was 5.00 am and just getting light. I'm sure I was asleep before my head hit the pillow.

I woke up at 11.00 am. It was a different world. The first thing I did was phone home and speak to my children. I didn't want them seeing television news reports of my Johnston Island incident without hearing from me first.

My stay at Honolulu was scheduled for two days and I was already several days behind my schedule. I was keen to get the Hawaii to Oakland leg over, as it was the longest water crossing of the flight. My schedule, however, didn't take into account the weather, and when I went to the meteorological office at Honolulu airport I was told to expect a 10 knot headwind which, after three days, should turn and give me a tailwind. So, here I was in Honolulu, waiting for the wind to drop and planning the next leg of my flight to the American mainland at Oakland, California. I wasn't going to risk flying with headwinds again.

My problems with fuel usage concerned me, and I used my time to figure out ways I could get better efficiency from my plane. It was vital that I get this problem solved, as the leg in front of me was slightly longer than the one I had just completed, and there was no Johnston Island to save me.

I had been ignored by Piper's Customer Service people on the several occasions I had written and faxed to them before I had left on my flight, so I knew I would get no joy from them. I spoke instead to several poeple in Australia and the consensus was to go with a zero headwind, and preferably a tailwind. I also spoke to a fellow woman pilot, Iris Critchell, who lives in Claremont, California,

I had met Iris, and her husband Howard, who she calls Critch, in Sydney the year before at a reception given in their honour by Nancy-Bird Walton, a famous Australian woman pioneer aviator. I had the pleasure of showing Iris and Critch some of Sydney. They ran a flying school in California and Iris had been an instructor pilot since the war. She had a lot of experience and I was pleased to have her advice and support.

Iris advised me to re-lean the fuel-to-air ratio every hour. This I would do by checking the exhaust gas temperature and leaning the mixture to peak temperature, then enriching it a little. But no one could give me any real reason for what seemed like excessive use of fuel on the leg to Johnston Island. Headwinds seemed to be the only major reason.

I haunted the meteorological office, but the forecast tail winds didn't materialise and, as though the Island's gods wanted to show me who was boss, the headwinds increased to 18 knots. I was stuck.

My time at Honolulu was a bitter-sweet time. Perhaps I should have realised the night before when I found myself going through United States Customs at the International terminal. It was in this place, ten years before, I first met Neville Kennard, the man who became my husband, fathered my second child and, in many ways, became the catalyst for my voyage. We were due to be divorced in only a few days time, our case being scheduled in the Family Law Court in Sydney.

When I first met Neville I was going to Washington D.C. to meet the man I thought I would marry and have a vacation with him in the United States. We intended to take in the snowfields of Aspen, Colorado, and so I had packed my snow skis. There was a man off my flight from Australia who was going through customs at the same time and who also carried skis. I remember thinking that he looked something like a country squire.

I could see that he had a certain gentleness, which is a quality I have always admired in men. We smiled and started talking about skiing. He was going to Aspen with his two sons from his first marriage. We had coffee together at the airport lounge and talked a little of where we lived in Sydney and what we did in our lives. We discovered that we also both shared a love of flying. Neville gave me his address in Aspen and invited me and my friend to visit him there.

The vacation was a disaster and my 'friend' soon showed me that marriage was not among his intentions. I did go to visit Neville while we were at Aspen, but he had gone away for two days. I left a note to say that I had called in.

I returned to Sydney rather confused about my now broken relationship, and finally went through the phase of '... well, who needs them anyway ...'

Obviously, Neville thought I did. He began a very slow courtship. I didn't want to commit to any sort of relationship at that stage. My mother was very ill at the time with the disease which eventually killed her, and I was looking after her, working, and looking after my daughter, Mimi. I kept Neville at arm's length. We would occasionally go out for dinner and really enjoy ourselves, laugh a lot and have a great time, and then on the way home we would develop a silly argument which seemed to take the shine off the night.

Our relationship was always tempestuous, and we seemed to go through a lot of sandpapering of each other, but our relationship developed and, after four years our son James was born and we were married.

There were many times when we split up and came back together again. And, as I explained earlier, it was after one particularly harrowing parting for me, when I hit the rock bottom of depression, that I decided that I would make something of my life and do what I wanted to do. It was in that time that the genesis of my flight was born. I knew that a destiny waited for me, and to realise it I only had to reach for it. It sounds simple to say, but it wasn't easy. I had to learn to stand in the sun as myself and not be in someone's shadow. I knew that was what I had to do, but I didn't then know how to do it.

The bitter times I had at Honolulu were not only as a result of waiting for the winds to abate; I was also constantly having to move. Honolulu is a major tourist destination and, because my original booking at the Sheraton was for only two nights, I had to book into other hotels, and at those times when I couldn't make a hotel booking I was blessed to find the support of members of the 99s, who lovingly took me into their homes. There were times in Honolulu when I felt like a waif.

The 99s are an international organisation of women pilots, formed in 1929 with 99 members. Amelia Earhart was a driving force behind its formation and the founding president of the organisation. The group was formed following the 'Powder Puff Derby', a race for women pilots across continental America. Back then, male pilots had their own races and women were kept out of their competitions — and their record books. When women started holding their own race, the press, reflecting the male dominant attitude in society of that time, dubbed it the 'Powder Puff Derby'.

It was against such attitudes that pioneer women pilots, like Amelia, set out to do their own thing — and flying around the world was not something many people had accomplished; she aimed to be the first woman. I think Amelia would have set out to discover new heights to climb, no matter what the attitude of the era. I also think that the attitude of the times she lived in helped shape the particular heights she chose to conquer.

Amelia had suffered in Honolulu. Initially, she had planned her around the world flight to go from east to west, departing from Oakland on St Patrick's Day, 17 March, in her twin-engined Lockheed Electra with a crew of three, bound for Honolulu. They arrived safely and at dawn, two days later, were taking off from Luke field when the aircraft ground looped and the landing gear partially collapsed. The damage was so great the plane had to be shipped back to the mainland to be rebuilt.

Amelia took off again from Oakland on 20 May, but this time heading from west to east. In Miami she described it as the long way to California. She disappeared after leaving Lae, on the home stretch.

People have asked me if my journey held more risk than Amelia's. Comparisons are really odious, particularly this one. She had a twin-engined plane and a navigator for company. I was alone in a single-engined plane with more sophisticated navigation equipment. Fortunately I completed my journey; Amelia Earhart, unfortunately, did not. I do know, however, that without her inspiration I would not have attempted the flight.

While contemplating my longest ocean crossing, I also phoned Fay Gillis-Wells at her home in Virginia in the United States. She was a charter member of the 99s and had been President Johnson's Press Secretary. She was very supportive of my flight and gave me a lot of encouragement.

When the meteorological office at Honolulu continued to give me no great joy, I decided to use my time there constructively. I had two extra fuel tanks fitted, which would give me plenty of reserve fuel. Air Service Corporation fitted two 18 gallon tanks on top of the auxiliary tanks 3 and 4 behind me. The new tanks were both gravity fed into tank No 2 beside me, and bypassed the auxiliary tank selector valve, which allowed me to switch to any auxiliary tank from which I wished to draw fuel.

My ten days of waiting on the Hawaiian Islands also held many joyous times for me. It was here that the major sponsor I needed to make myself financially secure, or almost so, introduced himself to me. Sherry Stumm in Sydney negotiated the terms of the contract, with the help of Nev. The news of a major sponsor was a godsend. When I heard the news I jumped for joy. Until then the financial aspects of my flight were pretty disastrous. I had left Australia in the belief that the necessary money would be forthcoming. Sherry's news confirmed my faith.

My new sponsors were Slender You Southpac Pty Ltd, the company that had the South Pacific rights for the Slender You slimming machines. I thought to myself that by the time this flight ended I would certainly be slim enough to advertise their products. Weight has never been a great problem for me, but so far on this flight I had lost over 4 kilograms.

Sherry Stumm told me that arrangements were being made for my plane to be painted in the sponsor's colours when I reached Oakland, and that two executives would fly from Australia to meet me there.

I had yet another reason to leave Honolulu as soon as I could, but still the winds held. I was learning the art of patience, of going with the flow, and in many ways I had to stop trying to control my life totally and just take each day as it came, whether it meant moving my base to another hotel, or when that wasn't possible, taking up the generous offers from the 99s.

I went from the Sheraton to the Holiday Inn for three nights, then to the home of Lois and David Luehring for a night. Lois is a 99, and David works with the Federal Aviation Authority as an accident investigator. We joked that he may have been called out to investigate me if I hadn't risked the landing at Johnston Island.

During my stay at the Holiday Inn I met an expatriate Australian, David Bell, the General Manager of the Bose Corporation in New Hampshire. I struck up a great friendship with him, Karen his wife, and daughter, Summer Jane. Summer Jane was 12, and reminded me very much of my daughter Mimi. The Bell family adopted me during their stay and helped me appreciate the tourist aspects of the island of Oahu.

I had met David through the Bose representative in Honolulu, George Mallaliou, and his wife, Joanie. George is almost blind and Joanie is his eyes. They are fascinating people with a real zest for life that gave me a boost when I needed it most. It had been organised from Sydney for Bose to give me a special noise-cancelling headset of the type that was used by Jeana Yeager and Burt Rutan in their record-breaking flight in 'The Voyager'. They minimise fatigue and ear damage, and my little plane was mighty noisy without them.

After eight days at Honolulu I knew I had to go. I decided

to improve my chances by flying to Hilo on the island of Hawaii, the most easterly of the Hawaiian Islands and 30 minutes flying time closer to the mainland. I spoke to Jerry Henry, the Chief Flying Instructor at Associated Aviation Activities, who told me he thought the weather was lousy for flying over water, even though Hilo was only a one and a half hour flight from Honolulu.

After some self-dialogue which went back and forth between self-accusations of being a 'scaredy cat' and self-congratulations for being wise and taking in the sound advice of experts, I decided that I would put off my departure for one day and would definitely go the next day. I told Jerry, who thought it a smart decision and helped me with standard instrument procedures for the next day.

My last night at Honolulu was spent at a cheap hotel near the airport. It was 22 August, the date my divorce from Neville was due to be heard. I had to know what happened, so I phoned him.

'Are we divorced?' I asked.

'No', he told me. Pause. 'I stopped the action.'

Good heavens, I thought, what's going on? 'Why?' I asked, very curious to know the answer.

His reply stunned me. 'Er, um, er, I didn't think it was appropriate ...' A long pause. 'How do you feel about it?'

Gosh, I wasn't too sure how I felt. His news had me on the hop. My feelings started to express themselves, anyway. 'I'm sort of pleased. Why?'

Our conversation went on for a short while and didn't seem to be getting anywhere. It was difficult to talk about this particular subject on the phone. I suggested that we continue it at a more appropriate time. Neville agreed and hung up.

I wasn't exactly sure how I felt about it. I'd picked up the phone expecting to hear that I was a 'single' mother again. Neville's news had knocked that expectation for a six, or into the outfield, depending on whether you follow cricket or baseball. In fact, it had knocked it into the bleachers. The more I thought about it, the more pleased I became. The door to our relationship, which had first opened on this island, was still open. There was still a chance that the love we had for each other could surmount the obstacles and bring forward the promise it always held. There was a chance we could get back together. And even while I was thinking these thoughts, I began to feel free of Neville. 'Perhaps this freedom is what love really is,' I thought as I drifted into sleep.

I took off for Hilo the next morning with the headwind down to only 7 knots and feeling considerably lighter in myself, too.

6: HONOLULU TO OAKLAND

The flight to Hilo was only a short hop and I landed early in the morning. I was greeted, as I clambered out of my plane, by a local pilot who did tourist flights over the island and also ran a tour bus. It was a quiet time for him so he volunteered to give me a tour of the island. I readily accepted. He was ideally suited to be a tour guide as he loved his subject and could (and did) talk about it at great length.

The island of Hawaii is dotted with the cores of extinct volcanoes, as are most of the Pacific Islands, but it has the distinction of having one very active volcano which has lava flowing into the sea. This adds to a tourist's sense of adventure. My guide took me to the lava flows, which are right alongside luxuriant tropical flowers and growth.

He warned me not to break off any of the old rock lava to take with me. It seems the Hawaiian goddess Pele is jealous of her lava and people who take it from the island run into bad times. They have particular problems with fires. It is not unknown for people to send rock lava back to the island with requests for it to be placed in its exact spot. He also told me that Pele is often seen around the island as an old woman with a white dog, sometimes sitting on the verandah or porch of a house, and when people look again, the woman and dog have disappeared — along with the house.

My impromptu guide dropped me at my room for the night, in a bed and breakfast establishment run by Sally and Dan Keglar, in their home. They greeted me with obvious warmth and sincerity and Sally took me upstairs to my bedroom, which looked out over the western sea. I had a wonderful view of the tropical sunset.

My stay at Hilo was calm and peaceful. The agitation that I had experienced waiting for the winds to drop at Honolulu left me, even when I found out the forecast for my take-off the next afternoon was for a 7 knot headwind. But this time I knew that, with the extra tanks I had had fitted, I would make it to Oakland.

That evening I strolled to the nearby village of Hilo and had a lovely meal in a local restaurant. As I sat by myself, I mused over the news that Neville had given me. Each time I thought about him I felt freer of him. Somehow my attachment to him was gone, and although I knew our relationship now had a better chance to succeed, I was not sure how it would work out. Back at my room I had a wonderful sleep and the next morning did justice to the bountiful breakfast Sally had made for me. I then spent the morning relaxing. I planned my departure for 4.00 pm that afternoon and ordered a cab to pick me up at 2.00 pm.

I waited for an hour and a half for the cab, and I was desperate. I phoned the cab company just as my cab arrived. The driver was a large island woman with very long hair and extra long fingernails. She had that wonderful island attitude to life that frustrates so many westerners who can't exist without schedules to keep. 'Don't worry, we'll get you there okay,' she told me. But at that stage I wasn't being philosophical about the attitudes to life of people of different cultures. I just wanted to get my plane into the air and get on my way to Oakland.

It was 5.00 pm as my little plane, laden with fuel, lifted off the Hilo airstrip. I wanted to take off in daylight because of my heavy load, and I also wanted to land in daylight. The only way to do this was to fly overnight. The weight of the extra fuel gave me a very slow climb rate of about 100 feet per minute. My usual climb rate was between 500 and 1000 feet per minute. So I had a slow climb to my cruising altitude of 8000 feet, even though I had sent all unnecessary luggage to Oakland by air cargo from Honolulu and about all I had kept on board was my life raft and other emergency equipment. This was the heaviest I'd ever been, about twenty per cent over gross take-off weight.

Cumulus powder puffs were scattered around the sky as I climbed north-east to the waiting night.

An hour out of Hilo I switched from my main tanks to No 2 Auxiliary. I checked my instruments and, lo and behold, I found that the Omega was working. I was delighted. It was a perfect night and I felt good.

I have never discounted the possibility that UFOs exist, although I have had no direct experience of any, that is until that night, when I noticed a bright green glow, away in the distance to my right, and seemingly coming out of the ocean. 'Good God, it's a UFO!' *Don't panic girl, there's no point in losing your cool. It's a long way off and you don't know that it's harmful.* The butterflies in my stomach didn't seem all that persuaded by the assurance. The glow grew larger and began rising out of the ocean. I couldn't take my eyes off it. It then seemed to just appear above the horizon, and I saw it for what it truly was — the waning moon, rising to soon disappear and leave me to the dark night. *Wow! It's the moon.* Very clever, I said, and then, to take my attention away from it, I checked my instruments again. The Omega was still working.

I suppose in a way I was disappointed that it wasn't a UFO. Somehow or other I felt cheated that I didn't have my very own close encounter, but then, on the other hand, I was really very pleased that it was only the moon, and I was comforted by its presence for the short time it was in the sky.

I was making good time and the headwind was dropping. I used my radio to contact my friend Tony Vacarella at the Overseas Telecommunications (OTC) back in Australia, and through him I organised a phone call to my friend Iris

Critchell in Claremont. Iris was going to fly her own plane from Claremont to meet me at Oakland and I needed to tell her that I was running an hour behind schedule. It was good to have the opportunity to talk to two friends and I didn't seem so isolated in my little cabin far from anywhere — a small speck in the cosmos.

Except for the nose tank, each of my auxiliary tanks was fitted with a visual gauge, which is a clear plastic tube on the outside of the tank. It allowed me to check visually the volume of fuel in each tank. I turned to my right to check No 2 and saw that it had a lot of fuel in it. This was odd. I checked my watch and did a quick mental calculation, which confirmed that there was more fuel in No 2 than there should have been. I checked again. It had one and a half hours extra fuel. I was puzzled.

I kept checking it as I flew and it was definitely using less fuel than any normal rate of fuel usage. I wondered if I might have been getting some heavenly assistance. As the fuel level neared the bottom of the gauge I followed normal practice and switched to No 1 tank.

Within seconds the engine coughed and stopped. There was absolute silence. Suddenly there was no forward pull and my speed rapidly decreased. My indicated air speed fell from 130 to 75 knots. I reluctantly pushed the nose forward to gather speed and prevent a stall. I had dropped to 7000 feet and was losing height quickly. I maintained 85 knots indicated. This was the best glide speed and I could see that I was getting closer to the ocean. The electric boost pump was on and I switched back to No 2 tank. Nothing happened. The ocean was closer and fear was tapping out its ugly dance up my spine.

Okay Gaby, switch back to the mains. I closed the throttle to stop air being sucked into the fuel system and enriched the mixture, then gently opened the throttle again. The engine came back to life. I opened the throttle to climb power and the plane regained its height and normal cruise conditions. I don't think I breathed again until I reached 8000 feet.

As soon as things were back to normal I began shaking with fear. A trembling seizure came over me. When the crisis hit I had fought the onrush of panic and had done what I had to do. All my years of training had come to the fore. The alternative was at best a ditching at sea with the remote possibility of rescue. At worst, and most likely in the dark, was death. Whatever resources I drew upon were deep within me and I was thankful that I had them. But now, I was overcome with emotion.

Gradually the trembling left me and I knew I had to make a decision. Turn back or go on. I was about five hours out of Hilo and well into the flight. The mains didn't have enough fuel to get me back to Hilo, or on to Oakland. Either way, I would have to use fuel from the auxiliaries. I decided to go on.

Once I made the decision the doubt and fear started to dissipate. My old friend came back. *All right girl, you know that when you've got the main tanks selected everything is working.* Yes, yes, I did know that. *Well, you've got to work out how to get fuel from your auxiliaries.* Yes. Yes. *And it worked okay on No 2.* Well, it did and it didn't, because No 2 took so long to empty. Why was that? *Now you're using your noggin.* Thank you very much. *Don't get feisty, we're in this together, remember.*

I hoped the breathing space of the fuel in the mains would give me time to work out this problem. There must be a logical explanation. I would find it. I relaxed a little. This was turning out to be the longest night of my life and I was living every second of it.

I've got to figure this out. *Right on Gaby!* Okay, what should I do? *You're the pilot, Gaby.* It might sound like flippant banter now, or even unbalanced, but the self-dialogue I conducted within my head helped me to get through.

I'm the pilot. That means I'm in charge. That means I know what to do to fix this problem? *Now you're talking.* Okay, I haven't had this problem before. *Nope.* And the only thing that has changed are the two new tanks I had fitted in Honolulu. So maybe the problem has to do with them. *Could be.* You're a great help. *Okay Gaby, consider this. Those tanks are divorced from the other system, they only gravity feed into No 2 tank.* Well, it's No 2 that played up. Maybe there is something wrong with the venting. *C'mon Gab, are you for real?* Well, I think I should check anyway. I'll drain the new tanks into No 2 and then disconnect the whole new system. *How?* Don't you worry, young Gaby didn't go to school for nuthin'.

I drained the two new tanks into No 2 and, as soon as it was full, I used a screwdriver and pliers to disconnect the hoses, which I then plugged with rag to prevent fumes escaping. I duck taped over the ends and then replaced the original cap on No 2 tank.

I could smell vapour in the cabin. Fuel vapour was dangerous, because the fumes could addle the brain and it would only take one spark to blow me to kingdom come. I opened the storm window a fraction and put my nose to it. The air outside was wonderfully refreshing.

I knew I had to crawl back between the tanks and check the vent lines and the drainage taps in case something was happening in either of those areas. This meant I would have to leave the plane on auto-pilot and hope that I didn't get stuck crawling over the fuel lines in that restricted space. It also meant risking a drastic change to the plane's centre of gravity, necessitating a very quick return to my seat. I wasn't too sure I could do it. My night was full of dilemmas.

Dilemma or not it had to be done.

What goes in can come out.

With every muscle aching I crawled between the tanks, checking the vent hoses along the sides and making sure there were no kinks in the hoses. I then used a ruler to stretch out to the drain taps to make sure they were closed. Everything checked out.

When I slid back into my seat I thought hard. I couldn't find the problem.

Let's be methodical about this, Gaby.

I happened to have a bottle of paper correction fluid

so I very clearly marked on the visual gauges of every tank the current level of fuel. I hoped this would let me see which tanks were emptying — and when. I think by now the process of elimination was starting to point to the culprit.

I've got a suggestion. Okay, what's the suggestion? *Why not try it on 2 again, there's fuel left in the tank.* That seemed reasonable. It was working on 2 and only cut out when I switched to 1.

I knew I had to conserve the fuel in my main tanks so, reluctantly, I made an agreement with myself to give it a go. I prayed: 'Dear God, help me. I must get back to my kids — please help me.' I gained confidence from the prayer. I turned on the electric boost pump and selected No 2 auxiliary, and then turned off the main tanks. I stopped breathing and counted.

At about 10 seconds, when I thought I had made it, I squeezed in another prayer: 'Please feed to the engine.' Four more seconds went by — then the engine sputtered. I quickly turned back to the mains. Shit, what is going on? A queasy feeling spread out from my stomach area and took over me. I broke into a cold sweat and began to feel nauseous.

I knew I would have to try it again. *Okay, take a deep breath and calm down. You can do it, you know you can. You're not destined to go down into the sea. You know that, surely you do.* I took several deep breaths and calmed myself enough to try again. 'God ... Please ...' My only prayer.

My hand shook as I followed the same procedure, though this time I made a change. Instead of selecting No 2, I selected No 3. I counted.

... 28 ... 29 ... 30 ... 31 ... 32 ... 33 ... 34 ... 35 ... 36 ... 37 ... 40 ... I started breathing ... 45 ... 50. The engine held. It was getting fuel. I was in ecstasy. I had three more hours of flying in No 3 tank. Hallelujah! 'Thank you, God.' *Amen.*

I relaxed a little. I decided to check the marks on the gauges. *Oh, oh, Gaby, we're in trouble.* Fuel was draining from 3 and 2 and faster out of 2. I checked the selector valve. The pointer was exactly where it was meant to be.

Right on No 3. The suspicion that had been working away at me was confirmed. For some reason the selector was draining out of two tanks at once. I remembered the words of the people in Sydney who installed it. 'It is impossible to select more than one tank at a time!' Well, here it was, the impossible was happening before my eyes.

I watched, horrified, as No 2 drained and the engine cut out again. I wondered how much of this I could stand. I was already well past my breaking point.

The boost pump was still on, so I enriched the mixture, turned to the right main and turned the auxiliary fuel off. The engine caught. I had to do it again. I had to find some way to get the auxiliary fuel to the engine and fiddle with that selector valve handle. How I wished that I had retained the United States registration so I would have paid a lot less money for a simple system that fed the one auxiliary into the mains. I wondered if my sense of patriotism was going to cost me more than money.

I steeled myself to go through it all again. I was past palpitations, almost to the point of resignation. *Don't be silly, Gaby. It can be done and you can do it.* I hoped my bravado would be borne out.

This time I switched to No 4. I waited, not even bothering to count. The engine purred ... and purred ... and purred. What a sweet sound. I checked the mark on the No 4 visual gauge. It was draining normally. I checked the other tanks — nothing was happening. I let out a loud 'Phew!' I checked my fuel situation. I had two hours left in each wing tank and three hours in No 4. All up I had seven hours of fuel which I knew I could access. I checked my position. I was five hours out of Oakland. I was safe. 'Thanks,' I told my God.

Just then a different voice said, 'It's all right, you can turn the boost pump off now.' This was not me talking to myself this time. I may not have had a close encounter with a UFO that night, but I did have a close encounter with some form of protection and guidance that I can only speculate on.

The night now had a mystical quality and, as I looked out of the window, I saw one of the world's greatest sights appear before me as a confirmation of this — a perfect sunrise, appearing golden above a bank of fog which stretched to the horizon. Its appearance marked the end of an ordeal during which I was tested to see if I had the right stuff to make it through. As I sat spellbound in that golden early morning glow, I felt as though I had passed through some rite of initiation and had finally earned my wings. I was in a state of euphoria, no doubt brought on by all the adrenalin that had pumped through my body, and the sheer delight in having made it through.

The only thing that concerned me now was the fog bank. I had been in San Francisco before and knew how the Pacific fog banks could stay for hours into the day before dispersing. But my good fortune stayed with me and my concern dissipated with the fog, about a mile off-shore. I was later told that it was the first time for weeks that the fog had burnt off in Oakland before noon.

The Golden Gate Bridge was glistening in the sun and the sight added to my emotions. I cried and cried.

San Francisco Control soon changed that. I was in their airspace and they directed me to approach Oakland from the south. This meant I had to deviate from my direct track, taking extra time, and I swore. 'Bloody Hell! Don't they know what I've just been through!' Fortunately, my subconscious responded, pulling my ego into line. *No Gab, they don't.*

I was warmly greeted as I moved into Oakland Control. My friend Iris Critchell was well-known to the controllers and she had asked them to give me all the assistance they could. They certainly did that, and the warmth of their greeting accentuated my need to step onto land again. I taxied to a stop fifteen and a half hours from Hilo, and I rushed to open my door. Two things were waiting for me that I desperately wanted to see and touch: my friend Iris, a fellow human being, and something that connects all humans — terra firma.

I don't know if Iris or I was more excited. We hugged

and embraced for what seemed like hours. A news crew was on the tarmac and I thought of the mess I must look. My pants were loose and just about falling off — I hadn't even thought to pull them back on properly. I had used the porta potty a few times on that leg from Hilo, and, with so much going on, I hadn't wanted to be bothered fiddling with my pants.

I gave a brief interview and headed straight to the waiting room with Iris. It was set up for a full press conference. I went in and made a bee line for the washroom. The difference a wash, some make up, a hair brush and a change of clothes can make is amazing. When I walked back out to meet the press I felt like a new person and much better equipped to handle the questions. It was the biggest press conference so far, with Channel 9 represented along with local press and wire services.

Iris took me to the home of Audrey and Tom Yeandle, which was a short distance from the airport. It was to be my home for the next two days. Audrey is a fellow 99. I shared a sandwich with Iris and Audrey and toddled off to bed. I slept deeply and woke in time to freshen up and join Audrey and Tom for dinner.

My stay with them was idyllic. Tom is a builder and his house is beautiful, no doubt because of Audrey's influence. I particularly loved their swimming pool, which was built in the style of a roman bath.

Their housemaid seemed to appreciate my need to rest after the hectic flight and was only too happy to bring me anything I wanted. What I wanted most was some clean clothes.

I spent most of the time at the Yeandle's home in the pool. It was ironic. Not too long ago I could have been swimming in a much bigger pool, the Pacific.

My plane was being serviced and receiving a coat of paint in my new sponsor's colours. I had time to catch up with the rest of the world. Channel 9 had booked me into the Huntington in San Francisco and for the next two days Iris and I behaved like tourists. It was good to get out on the streets amongst ordinary people. We did the stores, the cable cars, Fishermen's Wharf, views of Alcatraz prison. We were like two young girls — having fun and enjoying life.

Iris was due to return to her home in Claremont, so the next day we booked into the Oakland Hilton. My sponsors were booked in there, too, and, as it was much closer to the airport and my plane, I was happy to be there. The next morning we met my sponsors, Marcel Noe and Keshor Girdhar. They were both gentlemen, but determined to enjoy their time. They immediately invited Iris and me to a champagne breakfast. We looked at each other. Heck, why not? We were still two girls having fun.

I saw Iris off at the airport the next morning. How blessed I was to have such friends I thought as her Cessna 172 climbed on its southward journey home.

I checked out the progress on my plane. It would soon be finished and I would take off again on my journey. While there I asked a man who had been fitting ferry tanks for Pacific flights for twenty-five years to look at my selector valve. We decided that the indicator and handle were slipping. I would probably only select correctly by moving it in a clockwise direction.

I saw more of the sights of Oakland and San Francisco with my sponsors, but they couldn't stay forever, and I said goodbye to them at the International airport. My plane was ready and I had no further reason to stay.

When I got to Oakland airport I filed my flight plan and got into my plane. I was ready to take off, but I just couldn't. I didn't feel good about it. I felt really scared. The sweat just poured down my face and I thought, no, I can't do it. I was already cleared to leave Oakland. So I cancelled my flight — in a very professional way. I didn't want them to know that I was scared to death. I just said: 'I'm cancelling my flight plan, so I won't be going.'

I got out and walked around my plane and thought, oh my God, this is terrible, I'm chickening out. But my friend came back to help me through. *Well, that's all right, you can chicken out, Gab.* That's right, I can. So I just let myself chicken out, which was really good.

The next day I climbed into the aeroplane and I couldn't do it again. *This is serious, Gaby.* You don't have to remind me. I know it. It was serious and I didn't know what I could do about it. I had been up at 3.00 am to get to the airport and get everything ready to put in the plane and it was about 5.30 am and first light, and I just couldn't go. So I shut the engine down. *It's not the end of the world, girl. Be kind to yourself and do what you can to make it easier.* Thanks, maybe I can make it easier for myself. But how?

I had to work though this one on my own. Gosh, I'd already been through armed troops and engine failures, and now this — making it easy on myself seemed very sensible.

My plan was for another long flight of 11 hours. Okay, I could make that shorter. A shorther distance would mean less strain on me. That's what I'll do and I'll go in daylight. I'll leave at 10 am when the sun is out and the weather is good. That will be much better than taking off in this heavy fog. I had made a decision, and as I walked to the flight office to cancel my flight plan, the sun broke through the fog.

I sat in the little waiting room at the airport and it was sunny and lovely. I was really, really scared and I wondered what I was going to do. I could just get on a Jumbo jet and go home and forget all about it. I could sell my aeroplane here in Oakland.

But I couldn't be let off the hook so easily. *C'mon Gab, you know you couldn't do that. Admit it girl, that's not a viable alternative. You'll need more creative thinking. Go through all the possibilities.*

I had to admit that I didn't really want to abandon my trip now, just when I'd completed the most dangerous part across the Pacific, and the people in Amelia Earhart's home town of Atchison, Kansas, were waiting to show me Amelia's home. I wasn't going to abandon my trip. I would find a way. Going through the possibilities, including abandoning the flight, helped me get through it all.

It was a Sunday morning and there was a television set in the room, which was on. I hadn't noticed it before and, I suppose in an effort to avoid having to face myself, I turned to watch it. I don't know about chance, or synchronicity, I personally believe in divine providence. My experience of a voice telling me it was okay to turn the fuel boost pump off was a reinforcement of my belief, and as I was sitting in that little office, I received another reinforcement.

There was a motivational, inspirational growth and awareness program on the television and they were talking about fear and breaking through your barrier of fear and how you've got to face it to break through it. I knew that. *You know, this is what you're all about, Gab, breaking though the fear. Are you going to break through or are you going to sit back and not do it and hate yourself and feel bad and go home.* Gee, that's a bit tough. *Well, Gaby, we're in this together and we've got to get out. Don't worry, you're still breathing. You're alive, Gab.*

I thought, okay, that's true, I am living. I had always believed that we only really live when we have the courage to face our fears. That's really living. I knew I had a choice right at that moment. I could choose to go on with my flight and live, or back out and die slowly. I'm going to do it. *Yeah, Gaby, let's do it.*

I knew I had the support of many people, and the support of that force which had helped me to get to this point, and then arranged things so that I was sitting in the right room to watch the television program that I needed to see to get me going. I was reminded of a saying by Anatole France: 'Chance is the pseudonym God uses when he doesn't want to sign his own name.' For me the signature was very plain.

I filed my new flight plan and walked back to my plane and clambered in. The sweat poured down and that familiar shaking feeling started deep inside. I decided to ignore it and continued doing my pre-take-off check. I taxied to the departure point, holding onto myself and not giving in to the panic that hovered around the edges of my thoughts and wanted me to turn back. I knew I couldn't afford to let it in for even a moment.

I was cleared for take-off. As I pushed the throttle forward and heard the engine respond and felt the plane starting to roll, the panic decreased. When I reached the lift-off, I felt it lift, too. I was free.

I was in the air and actually flying again. I was in control. I couldn't believe that only a short while ago I was on the ground absolutely crippled with fear. I had been virtually unable to move. To get through that barrier I had had to contain the fear, realise it was using me up, instead of me using it to help me.

7: OAKLAND TO MEMPHIS

Happy? *Happy, Gab.*

We were up in the sky again, my plane and me, heading
to Winslow, Arizona, the destination I had chosen to make
this leg of my trip easy. My plan had been to fly to Bartlesville,
Oklahoma, to meet up with an old friend and then have only
a short flight to Atchison, Kansas. My plane was glistening
in its new colours and I was in a happy mood. I used this
flight to check carefully my fuel selector switch and got it
to work successfully, but only by rotating the handle one
way. Unfortunately, I noticed that my Omega had broken
down again, and was now totally useless.

Originally, I had planned to fly from Oakland to Denver,
Colorado, to meet Jack Munroe, an old friend and business
associate of Neville's. Jack is about 70 and the oldest teenager
I know. He flies a yellow Buker biplane, which is kept in
immaculate condition. The delays in Oakland had thrown my
schedule to bits and Jack was determined to go to an antique
airshow in Bartlesville for the Labor Day weekend. I had
agreed therefore to change my plan and meet him there.

I was looking forward to seeing Jack, as his spontaneous
approach to life and his high spirits never failed to lift me,
too. However, I was not disappointed about putting our
meeting back a day by going to Winslow; I needed to nurture

myself to make the rest of my flight with confidence, and doing the shorter 6 hour hop to Winslow, instead of the 11 hour stretch to Bartlesville was comforting. Once, not all that long ago, I would have made pleasing other people more important than my own needs and I would have made that long flight to please Jack. As I reflected on this I realised I had come a long way.

Winslow is a small Arizona desert town between two Indian reservations for people of the Navaho and Pueblo tribes. It lies at an altitude of 1600 metres and is known for the woven and bead work Indian artefacts on sale there. It has interesting remnants of an old Spanish settlement. Tumbleweeds blew across my path as I landed my plane. I remember being awed by the stark beauty of the high plateau desert when I crossed America in 1987. I have flown over, and landed on deserts in Australia, but they are mostly at or below sea level and any height of 1600 metres in my country is in rugged mountains, and is usually verdant.

When I got out of the plane I arranged for it to be fuelled up and started talking to a group of Texans. I've got to admit that, after talking to them, I became a believer that Texans do do everything bigger. They were on vacation in a luxury coach. To be honest, it was a super-luxury coach. It was a big clipper bus that had been fitted out with every conceivable luxury, including a sauna and spa. Unfortunately, while they were on their way to Winslow the air-conditioning unit had caught fire and the bus was extensively damaged. They were waiting at the airport for the owner to arrive from Fort Worth. He did, in true Texan style — at the controls of his own British made pocket-rocket or small jet.

I had a talk with him at the Winslow airport. What a character — 70 years old and looking forty-five. He had flown the South Atlantic route as a United States Air Force pilot during the Second World War. This was the route I would be taking, so I was eager to gain his impressions. He gave me some sound advice and, like the picture I had in my mind of a Texan, he treated me like a lady. It was great. He also told me that I was the second Australian he had met on an around the world flight. The first was some years earlier in Fort Worth, and that particular Aussie was flying a helicopter. It was our own Dick Smith, who has achieved so many aviation firsts.

I remembered the kindness and support Dick had given me during the preparation for my flight. His magazine, *Australian Geographic,* was a sponsor and Dick had personally given me a lot of very practical advice. Perhaps the greatest lift he gave me was a gesture of confidence that I would successfully complete my flight around the world. Dick entrusted into my care a piece of fabric from the Southern Cross, the famous plane in which one of Australia's, and the world's, pioneer aviators, Sir Charles Kingsford Smith — or Smithy as he was affectionately known to Australians — had blazed many new aviation trails. By giving me that small piece of fabric, Dick had expressed his faith in my ability. He knew that I would return safely to give it back to him. I carried it with me at all times.

Dick has long championed the cause of civil pilots in Australia and, in recognition of his expertise and understanding of the problems involved, he was appointed Chairman of the Australian Civil Aviation Authority.

I booked into the local Great Western Hotel and enjoyed a tremendous steak for dinner. I enjoyed my sleep that night. The next morning I checked out early and caught a taxi to the airport.

As I was going over the plane it occurred to me that I was carrying far too much weight. When I'd ordered my plane to be refuelled the day before, I had forgotten that Winslow was at an altitude of 1600 metres. The thinner air at that altitude meant that my plane would need more runway to reach take-off speed. At sea level, with the weight I had on board, the 2000 metres of runway at Winslow would be more than adequate, but not at that height. And it was already quite warm at 7.00 am.

I went to the local fuel man and asked him if he would help me to siphon off 40 gallons, which he could have for his own use. He readily agreed. After that job was done, I prepared for take-off, leaning the fuel/air mixture during run-up checks. At my first attempt, I felt I didn't have enough runway left, so I aborted the take off.

I taxied back for my second attempt. I actually went beyond the runway, and turning the plane I clamped my feet firmly on the brakes. I gave myself one stage of flap to increase lift and then I slowly pushed the throttle forward. As the engine's power developed I held my feet on the brakes and then gently released them. The speed slowly built up. 20 knots ... 30 ... 40 ... 50 ... 60. I lifted off as the end of the runway slipped beneath me.

A heavy plane and high, hot air — what a combination! I was pleased that it was flat country.

It took a long time to climb to my 8000 feet cruising altitude. I didn't mind. I was looking forward to meeting Jack Munroe in Bartlesville later that day.

About 100 miles west of Tulsa, Oklahama, it was difficult to avoid the thick cloud and heavy rain. I was closed in and wished that I had filed an instrument flight plan, instead of the visual one. Yet when I had taken off from Winslow the sky had been perfect and the weather forecast had been clear. I realised then that I was flying in that pocket of America where Dorothy of The *Wizard of Oz* had lived, and it was an unexpected storm that launched her on her very entertaining journey. I hoped the presence of the clouds didn't mean I was about to be launched like Dorothy. I looked outside my plane to see if there were any flying barns passing me. I couldn't see too much of anything because of the heavy rain.

As I approached Bartlesville the heavy clouds degenerated into rain squalls and I could see well enough to make a visual landing.

As soon as I got out I went looking for Jack, and although I found his plane, there was no sign of him. The people at the local flight base operation were very kind to me and gave me a lift to the Phillips Hotel in town. Everyone at the Antique Air Show was apparently booked in there. They said I would probably find Jack in his room.

After I checked in and unpacked I made contact with
Jack. It was great to see him and we agreed to have dinner
that night. I went back to my room to rest and then later
had a wonderful time with Jack and his friend, a retired
airline jet captain, who was planning to fly his single-engine
plane across the North Atlantic. Not bad, I thought, and
hoped I could do the same at seventy six.

On Tuesday morning I stood under the hotel awning
with Jack, and a group of other pilots from the Air Show.
It was raining so heavily that I knew, before checking the
weather, that all flights out of Bartlesville were grounded.

I couldn't take off, yet I had to get to Atchison. People
there were expecting me and I had already delayed my arrival.
I had to find a way to get there. I left Jack and went to
the desk and, with the help of the friendly people there,
planned a route to Atchison.

I made my arrangements. A hotel car would take me
to Tulsa, where I was booked on a commercial flight to Kansas
City. Virgene Smolik, my dear new friend and fellow 99,
would drive from Atchison and meet me at Kansas City
airport, and we'd drive back to Atchison.

I kissed Jack goodbye. He was amazed at my new
arrangements and I left him standing there with the others,
stranded like gulls.

The short flight to Kansas City reminded me that this
is the country of terrible storms, where twisters can appear
out of a clear sky and demolish towns. I don't know how
the pilot of the 727 landed, the conditions were appalling.
The cloud was down to 50 feet, but we made it.

Virgene was waiting for me and we shared the short
drive to Atchison. We drove to Atchison airport where the
mayor had arranged a presentation for me, thinking I would
arrive by plane. I was over an hour late, but at last I was
in Amelia's home town.

Whenever I think back to my two days in Atchison the
tears come to my eyes. I knew that a reception was planned
for me, but I had no idea of the warmth that was to be
extended to me. Although my circuitous route had made me

so late, Mayor Dave Dennis had waited for me at the airport, along with a small group of media representatives, who were joined by some hardy local townsfolk. In a very touching ceremony, Dave Dennis told me that the city council had declared that day as 'Gaby Kennard Day'. He very graciously presented me with two keys to the city, one that I could keep and the other to carry with me on the rest of my world flight, but which was to be returned to the citizens of Atchison within a year.

This gesture was so important to me because, like Dick Smith's gesture, it showed a faith and belief that I was going to make it. Also, it was significant to me because it meant I was now carrying a reminder of Amelia Earhart.

I was thrilled, and yet bewildered that I should be so honoured. I didn't realise there was more to come. I was given a motorcade through the town, and although most of the people had given up waiting and gone home (I don't blame them. I'd have done the same!), it was exciting to be given this sort of welcome. I had to keep pinching myself to make sure that it was for me.

But one of the most moving and memorable moments for me was my visit to the home Amelia Earhart had grown up in. In the early years of this century, Atchison was a famous railroad town on the Missouri River. Amelia's father was a lawyer for the rail company, which was the same as the one in the song with the line: 'The Atchison, Topeka and Santa Fe'.

Atchison has some beautiful houses from that era which their owners lovingly maintain. There are many old two-storey homes with wonderful architectural additions such as gables, widow's walks, and the like. Beautiful stained glass windows adorn these houses, which reminded me of the Australian homes in the Federation style, although the American homes had wonderful, fancy, interior wood panelling.

Amelia's home, which is now a museum dedicated to her memory, is typical of the houses of the period and overlooks a sweeping bend in the river. I entered the house, with a sense of unease, and was shown through it. The furniture has been retained where possible and it must look very much like it did when Amelia and her sister Muriel ran through it in their bloomers.

The other people in the house became an intrusion to me and I found myself alone in Amelia's bedroom. I was in another world and could sense her presence in the room. It was so strong and I can remember looking out her bedroom window, across to the river, and imagining that she had just stepped out for five minutes and would soon return.

Poetry is definitely not my forte, but my feelings in that room were so strong, that I penned these words in respect.

Dear Friend,

I left this note in case you soon return and find that I could
*   wait no more*
I travelled through the years to find that you have gone
I spanned the ocean waves that took you in. I brought you
*   home.*
I sense that you are close
Yet I know that you have gone.

I heard your voice. You were playing in the yard.
But when I turned to look you were not there
And your voice became a whisper in the wind.

I left this note in case you soon return and find me gone
To treat once more with wind on waves
To soar above the earth like you.
I follow on the passage that you made
That path of truth you held to be.
The true in me is clearer now.

We'll fly together on another day, dear friend.

Gaby

Being in Amelia's house, the house of Amelia's childhood, brought back a flood of other memories for me — memories of my childhood. I know that my early life shaped me and gave me a strong sense of independence, which I learned to suppress, but which, through some self-transformation classes, I had been able to develop recently. I had finally learned to recognise it as a positive force in my life.

I spent most of my young life in a world of my own, a world that had no room for the violence and arguments between my mother and Thomas Dudley, who I thought of as my father. My mother was the aggressor in these fights, which were usually brought on by heavy drinking. I vividly remember the terror I felt as their arguments built up to physical violence and Thomas would plead with my mother to stop. She would attack him with whatever was available and I remember one night when she smashed a bottle over his head. He was bleeding and crying and I had to pick pieces of glass out of his scalp.

Although there were occasional happy times, a general pall of misery hung over that house. My mother was very unhappy and would often do cruel things to me for no apparent reason. She once sent me to school with string instead of ribbon in my hair. I think the reason for this was because I had lost the ribbon. She once boasted to me that when I was six weeks old and she was out walking with me in the pram she stopped to talk to a group of women and I started crying. She proudly told me that that was when I had my first hiding, because she was not going to be dictated to by a six-week-old child.

I found escape from my mother and her problems in the wonderful world of my imagination. Here, deep inside myself, cut off from the world outside, I was safe. I could build my 'real' world, from which I could draw the nurturing all children need. I liked being above the ground and had two favourite trees that I would climb. The tree in the yard had a tree house and in there I would pretend to fly. I knew that if I pushed my belly-button I could fly like the fairies who were my constant companions.

I loved fairies, I think because they could fly and because they represented a world of beauty and wonder — the type of world I wanted to live in. I set up a place in the garden which became my fairy garden and I would do everything I could to make it special — pretty, so the fairies would always come there. I had a litttle house in there that I would decorate. The roof came off, so I would play around, talking to the fairies and generally being in my own world.

My other tree was quite some distance away from the house, near the railway station, and I usually only visited it on nights when the moon was out. I would sneak out of the house. I loved walking in the moonlight to my tree, which I would climb and hide myself in. I would watch the trains pull into the station, and the commuters. Some would even walk beneath my tree, but they didn't know I was there. In fact, they didn't even look up. It wasn't until many years later that I realised why the burdens of adult life keep most people from looking up.

After I had had enough of tree sitting, I would let myself down and walk home, playing games in my head with the moonlight and the shadows. At home, I would quietly let myself in and sneak back to bed to dream of fairies and being above the ground.

I was about five years old at this time and ready for school. My mother had been raised as a member of the Catholic Church and, although very much a non-practising member by this time, she decided that I should also receive the benefits of a 'good' Catholic education. I'm afraid they sent me to the wrong school for that. I don't know why they chose the school they did, perhaps because it was the only decent one in the area, but the benefits of the faith I received there were enough to turn me against Catholicism and any organised religion. I hated school and was miserable.

But one especially happy event happened during those years. My sister Sheri was born when I was five. I adored her. When my mother bathed and changed Sheri, I bathed and changed my doll. It was the greatest time. We had lots of fun — my mother, my baby sister and me.

My mother showed her independence when I was seven and Sheri was two. She left Thomas and we moved from Melbourne to Sydney. She soon got a job as a barmaid at a hotel near Centennial Park. The job had accommodation provided, so my aunt looked after Sheri and I was placed at a boarding school run by the Marist Sisters at Woolwich. My earlier experience of Catholic schools was still fresh in my memory, so I was very reluctant to go. Fortunately the Marist nuns were wonderful and knew the value of a kind word and the comfort of a soft voice.

There were only three other boarders of my age, so it was still lonely and I would look forward to Sundays when I could go to my aunt's and play with Sheri and my cousins. I think I know how prisoners who are given day release must feel.

I reflected on all this at Amelia's house, and saw how I could have developed my independence from that sort of childhood.

I stayed in Atchison at a motel owned by a friend of Virgene Smolik, and I was given a reception at Amelia's home on my first night. I found it hard to accept that these people were there to follow me. There is no doubt that the people of America are warm and friendly and very sincere, especially those in the heartland of the country who are not influenced by the glitter of the large cities. If there is one characteristic above others that endears me to Americans it is their obvious pleasure in getting behind people who are setting out to achieve their goals in life. There is an infectious joy that says, 'Good on you, we know you can do it'.

The people of Atchison, and the 99s, demonstrated to me what people around the world can do when they decide they would rather be part of the solution than part of the problem. They have established a park, which they call the International Forest of Friendship. This is a park in which trees from every state of the United States, and from every country that is represented by a 99, have been planted. The walkways throughout the park have plaques set into them which commemorate the achievements of aviation pioneers and achievers.

On my second day in Atchison I was taken to the Forest of Friendship and given an escorted tour by its co-chairman, Joe Carrigan. Joe's obvious delight in the Forest and what it represents as a contribution from the people of this community towards world peace was infectious. I was reminded of a saying I had heard somewhere: Peace begins with you. Obviously the people of Atchison had taken that saying to heart and, instead of wishing for peace to happen, they had planted trees and commemorated those people who had helped to make the world more accessible and the realisation that we are all one people, closer.

I caught the essence of Joe's enthusiasm, especially when he showed me the names of some Australian aviation pioneers. I was shown plaques commemorating the achievements of Australians Sir Charles Kingsford Smith, Nancy-Bird Walton and Dick Smith. Nothing, however, prepared me for the surprise when Joe lovingly showed me a plaque on which my name was inscribed. I was overwhelmed by this gesture and the emotion was too much for me. I couldn't hold back my tears. To think that these people thought enough of what I was doing to pay me this tribute was very powerful. I really didn't have the words to express my gratitude to the citizens of Atchison for their courtesy and generosity of spirit. Thank you sometimes seems very inadequate.

My last night in Atchison was celebrated with a lovely dinner given in a beautiful old home, part of which is used as a restaurant and reception centre. It was a memorable finale to my visit. But my flight beckoned and the next day I had to continue on my journey.

My plane was still in Bartlesville and Virgene Smolik and her husband, Vince, decided they would fly me, along with their little dog, to Bartlesville. They love flying and have a single-engined Mooney, an oldish plane kept in top conditon.

It was a tight fit, but we made it to Bartlesville and, after posing for a lot of photographs for Virgene and Vince, and bidding them a grateful farewell, I took off for Memphis, Tennessee at 4.30 pm.

I was looking forward to arriving in Memphis. My plane was due for a service there and I was hopefully going to get my Omega Navigation System fixed. This would give me some time off and I intended to go and see Gracelands, to satisfy my curiosity, as Elvis was such a part of my early teenage years.

I made my final descent into Memphis International Airport as the sun was setting. It was 7.00 pm.

8: MEMPHIS TO PARAMARIBO

I was looking forward to my weekend stopover in Memphis, Tennessee. It was a name that was familiar to me from the 'heady' days of my teenage years, when Rock 'n' Roll was still young enough to be suspect by mature adults and so a natural attraction for teenagers trying to find their individuality and break away from the social bonds of the era that was passing.

The people of Memphis gave me a taste of real Southern hospitality. I was met at the airport by Nancy Miller, the Memphis Governor of the 99s. I was reminded again of how wonderful this organisation of women pilots was, and how they had really supported me on my flight. Nancy is a postal inspector with the United States Postal Service and lives out of town on a small acreage which she shares with her dog and horses, but she organised a small reception for me at the airport the next morning at 11 am.

I arranged to get out there early and get my plane booked in for a service and to get my Omega repaired. Memphis airport was the only place in the United States where I could have this done. It was a Friday and the service people at AMR Combs couldn't carry out all the necessary work on that day and told me they would complete it on Monday. This meant I would have to spend the weekend in

Memphis, which I didn't mind doing at all. I was getting closer to leaving the States and I must confess I wasn't looking forward to it. I knew I would have to face more water crossings and be among people who didn't speak my language.

The 'small' reception that Nancy had organised for me included the mayor of Memphis, who gave me the key to the city and a certificate granting me the status of honorary citizen. This was a further honour, which I found hard to accept. Somehow I felt like a fraud because I didn't think I lived up to the image these people had of a 'heroine' flying around the world.

Among the group at the airport to welcome me was an 85-year-old man who had pilot's licence 104. I think he and the people who flew in those early days when aeroplanes were still a rarity are the real heroes and heroines of the air. Members of the Memphis-Shelby Country Airport Authority were also at the reception and there I met Sandra Kelley who later took me to see Gracelands.

In many ways Gracelands was as I thought it would be — tacky but interesting. I still vividly recall the impact that Elvis Presley had on my peers and those a little older. I was living at the time with my mother, and her partner Laurie, who I thought was her husband. We lived at Seaforth and I was a student at Manly Girls High School. I had broken away from the Catholic education system after my mother took me to see a psychologist.

When I had been nine, and still at boarding school, I missed a home, and family life, terribly and so I asked my mother if I could go back to Melbourne and live with Thomas Dudley. Surprisingly, she readily agreed, I imagine because the fees were a heavy financial burden. For the next three years I lived in Melbourne and found myself back at my earlier, hated, school. Thomas was living with a woman called Fran, who had two daughters, one a lot older than me and one about my own age. I completed my primary schooling while living with them and then, as my mother seemed settled in with Laurie, I went back to Sydney.

My mother could see that I was abnormally shy and took me to see the educational psychologist, who suggested that the cloistered life in the Catholic schools had sheltered me from the world to the extent that I lacked normal aggression. He suggested that I go to a State school which would help me to mix in with other kids and become more assertive.

I was enrolled at Manly Girls High School and I took that opportunity to break with the Catholic church. I wasn't sure whether I believed in God or not, but I knew I couldn't subscribe to the God that had been portrayed to me in my formative years. I think I developed a strong dislike for any creed that claims to have the exclusive rights on salvation. I'm sure God is a lot more universal than that.

I had one more joust with the Catholic church, however, but I have to confess that my motivation was not inspired by any lofty purpose. It was much more mundane. He was the best-looking boy in the neighbourhood and I knew he was attracted to me, so I agreed to meet him one Sunday at Mass and then go to his nearby home and have morning tea with his family. His parents were very strong in the Church and went to Mass every day.

The morning tea went without a hitch until his mother let slip, in a very snide way, that it was a shame that Laurie and my mother weren't married. I was shocked, not so much at the news, but at the hypocritical way it was conveyed to me. On the way home the boy told me he didn't think it was a good idea for him to continue seeing me. He needn't have bothered.

I think my desire to break out of the repression of my childhood made me a natural to gravitate to Rock 'n' Roll. When I was fourteen I started going to the television studio which broadcast the Australian Bandstand program. I would go every Saturday and soon became pals with another regular, a girl called Maxine. We became so well known that Brian Henderson, the program's host, asked us to take turns at being hostess. It was very exciting at the time and a whole new world for me. I became reasonably well-known and was, as a result, given modelling assignments.

Another regular was a young man from Mosman called Richard Neville. He and I hit it off and became good friends. I would get up to all sorts of pranks with Richard and often we'd sneak off to Kings Cross, which was then the centre of Sydney's bohemian life, although it was going through the transition period that would turn it into the drug and prostitution captital of the South Pacific. As a fourteen-year-old I found it fun and daring to dance with Richard in such meccas of the high life as El Rocco and El Bongo — it was that period in history when all that was chic began with an 'El'. Richard was fun to be with and knew his way around.

My Rock 'n' Roll years were full of colour and excitement, but being at Gracelands and viewing the displays of Elvis's costumes from which the colour was fading, and seeing the plastic flowers arranged in the meditation garden in which he and members of his family are buried, repelled me. The feelings of nostalgia with which I had set out to see Gracelands were soon swamped in a miasma of nauseating commercialism.

I shook off the Gracelands blues at a small dinner party that evening at Nancy's home. One of her co-workers was transferring and she had eight people around for a farewell. It was good to be with ordinary people and listen in on the local gossip. One thing I have learned from my flight is that people enjoy a good gossip no matter what part of the world they live in. It's a good diversion.

On Sunday we went to Olive Branch airport, a short distance from Memphis. It's a small, popular, secondary airport used by a group of flying enthusiasts. I met many people, including Jimmie and Lean Hunt. Jimmie is the proud owner of a P51 Mustang, one of the fastest fighter planes in the Second World War. He calls his plane 'Contrary Mary' and took me up for a fly.

It was a thrilling experience to listen to the sound of the big rotary engine and propellor working in harmony to pull that plane through the air. Jim has an endorsement to do barrel rolls at 500 feet and he did several of them — what a thrill! It's a great plane and he's a wonderful pilot.

Gail Wardlow and her husband had an old Helio Courier Short Takeoff and Landing (STOL) plane which, because of special slats in the wings, can take off and land at speeds as low as 20 knots. They let me fly it, which was a very different experience. I also flew a Glassair R6, a very small, home-built plane which is made of composite material and is very light. Its aerodynamic design and light weight allows it to fly at 240 miles per hour.

That evening a group of us drove south of Memphis into the State of Alabama, which is very much as I imagined it to be. It is the deep south and we went there to enjoy some real southern hospitality. We drove to an old-fashioned cafe where I had catfish, which I thought very much preferable to alligator. The location could have been out of a movie set and the deep drawls of the local people added to the atmosphere.

The next morning Nancy drove me to the airport and I had to wait a little while for AMR Combs to finish the service. They were so nice to me, and refused to take payment. They also filled my arms with tee shirts, hats and scarves.

At 10.15 am I gathered speed to lift off at Memphis International ?nd set my course for Fort Lauderdale, Florida, my last stop on the North American mainland.

I found the people at Fort Lauderdale as friendly and as hospitable as those elsewhere in America, and they looked after me in grand style. I stayed one night with Alex Ewanchew, who met me at the airport on arrival. The second night I spent with Ellie Richenbach, who told me I would be going through part of the devil's triangle. My old companion was back. *Did she say the d-d-d-devil's triangle, Gab?*

I managed to make it to the airport without any apparent shaking. But when I climbed into my little plane I almost had a repeat of the hesitancy I went through at Oakland. Fortunately, I managed to control myself long enough to become airborne and once more experience the joy of flying.

The United States had been good to me. It was my

landfall after the most harrowing night of my life. I was honoured by the people, who had been extremely generous to me. It was no wonder I didn't want to leave. But I had my flight to finish and my own children to get home to. I was now missing them like crazy. Being in other people's homes brought on my homesickness. And I was missing the crazy Aussie accent. I needed to get on with it.

My flight path was down through the Caribbean to San Juan, Puerto Rico, and it was a good 7 hour flight over water. I was under the control of Miami Centre and it was good to hear the voice of the female flight service operator.

I moved through a tropical wave, which is an area of convectivity, that is of instability and storms, with the odd flash of lightning, so I skirted around the storm cells and through to lighter cloud. I breathed a sigh of relief when I broke through into better weather.

The Miami Centre controller had requested that I check in again when I reached a certain position, so I did. But I received no reply. I tried several times, and tried everything. It was certainly odd. It was like talking to a blank wall. Thoughts of the triangle crossed through my mind, but all looked clear outside, so I thought it best to wait and see, rather than worry.

I flew on course for about 20 minutes when over my VHF radio I heard two pilots talking: 'Maybe she's gone to the potty.'

I thought they were talking about me when all of a sudden, clear as a bell, I heard 'Victor Hotel Golf Kilo Foxtrot, I assume you're cruising 9000.'

Oh, Joy!!!

I calmly replied: 'Miami Center Victor Hotel Golf Kilo Foxtrot cruising 9000.'

I don't know still whether those pilots I overheard were referring to my friend back in Miami. If so, she had a long break, and even so that wouldn't explain why I couldn't raise anyone else. Perhaps I flew through an area which had freakish atmospheric conditions. Maybe it was the triangle. Maybe the girl did have a long break. Whatever, it remained a mystery to me, and I was quite happy to keep flying and not bother trying to solve it.

I did notice, over this area in particular, how rapidly the wind patterns would change. The Pacific had taught me the value of keeping an eye on the water as a gauge to the prevailing local weather, and that habit was second nature to me now. All through the Caribbean I noticed how the wind suddenly changed. One moment I was flying over choppy water with the chop indicating a strong nor' easterly, then a short time later the water was oily, indicating a dead calm. I used these visual clues to correct my plane for drift.

There was heavy cloud on my approach to San Juan, and I dodged a couple of storm cells as I brought my plane in for my first taste of the exotic Caribbean, although Puerto Rico is not among the most exotic places in the Caribbean. It is a protectorate of the United States and the local economy is not strong.

Several journalists were waiting for me and dropped me at my hotel. It wasn't very good, but at least it was a bed.

When I had been in Oakland I had talked one day to Iris Critchell and told her of my concern about the lack of accurate weather information I would be able to gather once I left the United States. She suggested that I make contact with an organisation called Weather Service Corporation, in Boston. They are a commercial weather information service which uses the latest satellite technology to advise their clients on world or local weather patterns. I thought that I might really need them, so I had phoned and organised with Bob Rice to make contact again in Puerto Rico. He was of great assistance to me and was able to reliably advise me of weather patterns affecting my route. In my mind Bob became 'Boston Bob', a voice at the end of the phone with whom I could make contact from places with exotic names and no reliable weather information. It was great to talk to him.

I phoned Bob that night, and he told me that Hurricane Hugo was very active and looked like it was heading in a north-westerly path that would bring it menacingly close to me. He advised me to change my flight plan. From Barbados I therefore planned to head on a more southerly course which would take me over Trinidad and bring me over mainland South America, further west than planned, earlier, keeping me out of the path of the hurricane. Nevertheless, Bob told me to keep in touch.

Although San Juan didn't reflect the affluence I saw and enjoyed on the mainland, the people shared the same generosity of spirit as the Americans. The next morning at the airport the local Flight Base Operator, handling agent, Juan Villefane, gave me free fuel, a thermos of coffee, and a cooler with ice and coke and some sandwiches. The Puerto Rican wing of the Air Force, who shared the same airport facilities, gave me caps and badges and other small mementos. What a treat to have a cold drink or hot coffee en route.

I took off at 8.30 am, heading for Barbados. The calypso beat seemed to affect my small plane and it was buffeted about the hot Caribbean skies. I thought of Harry Belafonte and the wonderful songs which introduced the world to the West Indies. I joined my plane and sang every calypso song I knew. We flew across the skies, over islands with exotic names like Martinique, to the haunting beat and melodies of 'Yellow Bird', 'Jamaica Farewell' and other toe tappers.

For most of the time visibility was poor, caused by a haze that was similar to flying through cloud, yet it was not cloud. It was spooky and I thought I was in a movie about the devil's triangle. I think my singing was really to keep my fear at bay.

I spent only one night at Barbados, even though I would really have liked to spend more time. Boston Bob advised me to get out of there, fast, as Hurricane Hugo was heading that way and it was 'one helluva storm'. Bob's prediction was chillingly accurate and his thoroughness saved my journey from disaster and my plane from becoming one more piece of twisted metal on the tarmac.

Pat Callender, the airport director, welcomed me on my arrival at Barbados and invited me into his office for a cold coke. He was a charming man and was halfway through giving me a potted history of the country when the door burst open and we were interrupted by a most neurotic character. He was an Austrian pilot, employed by one of the few air services in nearby Surinam. His job was to fly passengers from Paramaribo, that country's capital, to places like Barbados. His plane, an old light twin relied only on the pilot's navigation skill to get to where one wanted. It was in a bad state of repair and the only bit that worked, apart from the engine, was the compass. No wonder he was neurotic.

This pilot, who no doubt suffered incredible pressures through flying on his local knowledge and by the seat of his pants, complained to Pat that the airport staff were treating him badly and it was because he was white. Pat, who was black, looked at me with an exasperated expression and then patiently proceeded to pacify him.

This incident made me pleased that I had cancelled my original flight plan, which took in Paramaribo, where Amelia Earhart had landed. My friends in America had talked me out of it, because it was such a depressing and unstable place, and I would receive little help there. Apparently Surinam had been a Dutch colony and very much favoured as a tourist destination, but recently, tourists had stopped coming, money dried up and the country became stagnant, and then backward.

Pat was a cheerful man, who insisted on driving me to my hotel. We drove past rows of neat bungalows which looked like dolls houses sparkling in the sun.

The hotel was typical of a resort complex and could have been in any tropical location in the world, with outside tables with plastic grass umbrellas, lots of bars, and swimming pools. It seemed incongruous to me and completely out of character with the rest of the island.

As soon as I unpacked and had a quick wash, I phoned my children. They were always on my mind, and more so now that I was well into the flight, and well over my schedule. At every stop they were my first priority. I made lengthy calls. This constant contact with them helped me to keep my motivation to continue with the journey. Besides, Jamie's class at the North Sydney Demonstration School was studying each country I visited and I had made a commitment to them to mail them information from all the places I stopped.

My kids, God bless them, were great supporters of my flight and, even though they missed me and wanted me back, they wanted me to finish. Before I left Australia we had had many long talks about how important the flight was to me. I had had some criticism from sections of the Australian media who seemed to take the position that it was scandalous for a mother to go off on a 'jaunt' like mine and leave her children behind. *60 Minutes* had involved my sister Sheri in some of this, and for a short time our relationship was strained. Fortunately, the love Sheri and I have for each other is built on a solid foundation and it only took for us to sit down together and have a long talk for our relationship to return to normal and for Sheri to become a big supporter of my flight.

My attitude to my flight and my children was, in many ways, I suppose, formed by my experience of my own mother. I saw what sacrifice did to my mother. She chose to sacrifice her love for my father to live with Thomas Dudley and it made her a very unhappy person. She carried her unhappiness around like a sackcloth and took every opportunity to remind me of her martyrdom.

I'm not a great believer in the 'me' consciousness that was spawned in the '60s. But there is a big difference between being selfish and 'self-ing'. I don't think any person can fully contribute to society at any level, from that of the family, to the local community, to the nation, to the world, unless they're happy in themselves. And I believe that happiness comes when we are doing those things that we are good at doing, that we want to be doing.

I think, too, that it is important for all people, and particularly women, to know that they have equal opportunity to pursue their dreams of a useful life.

I knew that I would be more useful to my children if I pursued my dream of the flight, than if I took the easy way and was confortable but unfulfilled. Even if I never

mentioned it, and that was unlikely, they would have been
saddled with the burden of Mum's sacrifice and that would
be a terrible thing for them to live with. They would,
unconsciously, have spent the rest of their lives pleasing me
— instead of living their own lives.

Certainly, the risk of my death was a factor to be looked
at and weighed, yet the possibility of my dying on my flight
was the preferred alternative for everyone, than the slow
death of the spirit which would have occurred if I hadn't
done it. I believed strongly that I wasn't going to die on
this flight.

To do the flight, however, I had to face many fears, and
perhaps the greatest was the fear that my children would
develop a sense that they had been abandoned. The events
in my own life seemed to reinforce that possibility, and,
because of my own first hand experiences of abandonment
and rejection, I knew how to involve my children in what
I was doing so they wouldn't suffer the same fear.

The catalyst for my flight had been when Neville walked
out on me — abandoned me, rejected me. Somehow, in my
depression and my grief I found the spark of life that said:
'Enough! I want no more of this! I am worth more than this!'
I had spent many years listening to negativity and it had
stopped me doing many things. It wasn't going to happen
again.

After I phoned my children and had a long chat with
them, I phoned Boston Bob and was given a weather update.
As soon as I got through to him he told me, plainly, to get
out as soon as I could.

Pat picked me up from the hotel the next morning and
the local people gathered to say goodbye. I think Barbados
was the happiest place I had been in. The people just seem
to have an inner joy. I was genuinely disappointed to be
leaving.

My flight path took me south from Barbados, over
Trinidad to mainland South America at Venezuela near the
mouth of the Orinoco River. I then adjusted my course to
a south-easterly direction somewhat inland and paralleling

the coast. While I was heading south-east Hurricane Hugo was following its north-westerly path of destruction only 300 kilometres to my east. It was a very severe hurricane and was devastating everything in its path. Weather Service Corporation and Boston Bob had saved me from a closer encounter with an unbridled hurricane.

I had to adjust myself quickly to flying conditions over the jungle. My only previous experience of flying over jungle had been in Papua New Guinea, but here I was confronted by a continuous panorama of dense treetops. It was an alien world and I couldn't help speculating on what would happen to me if I was forced to land. It was a morass that would soon swallow me up.

My planned landing was to be at Cayenne in French Guiana, and my course took me over Guyana and Surinam. As I continued southwards I ran into very turbulent cumulus build up, which is common over equatorial jungle regions. I could see massive cloud formations ahead and decided to ask Zandery tower in Surinam to get a terminal area forecast for Cayenne. They told me that there was a nasty storm in the region and the cloud base was 3000 feet with the tops at 55,000 feet, and no break in between. I knew I would be in trouble if I went on.

I requested permission from Zandery tower to land there. 'Is it permissible?' I asked, remembering that this was a country run by a military junta. The controller was very friendly and told me it was not usual, but he would see what could be done. I descended to 2500 feet to become visual.

The controller came back with the good news that it was okay for me to land. Despite my best endeavours, and against all the advice of my American friends, here I was landing in Surinam. And all I knew about Zandery airport was that it was surrounded by jungle and an hour and a half's drive to Paramaribo. As I touched down I wondered how I would get from the airport to the town.

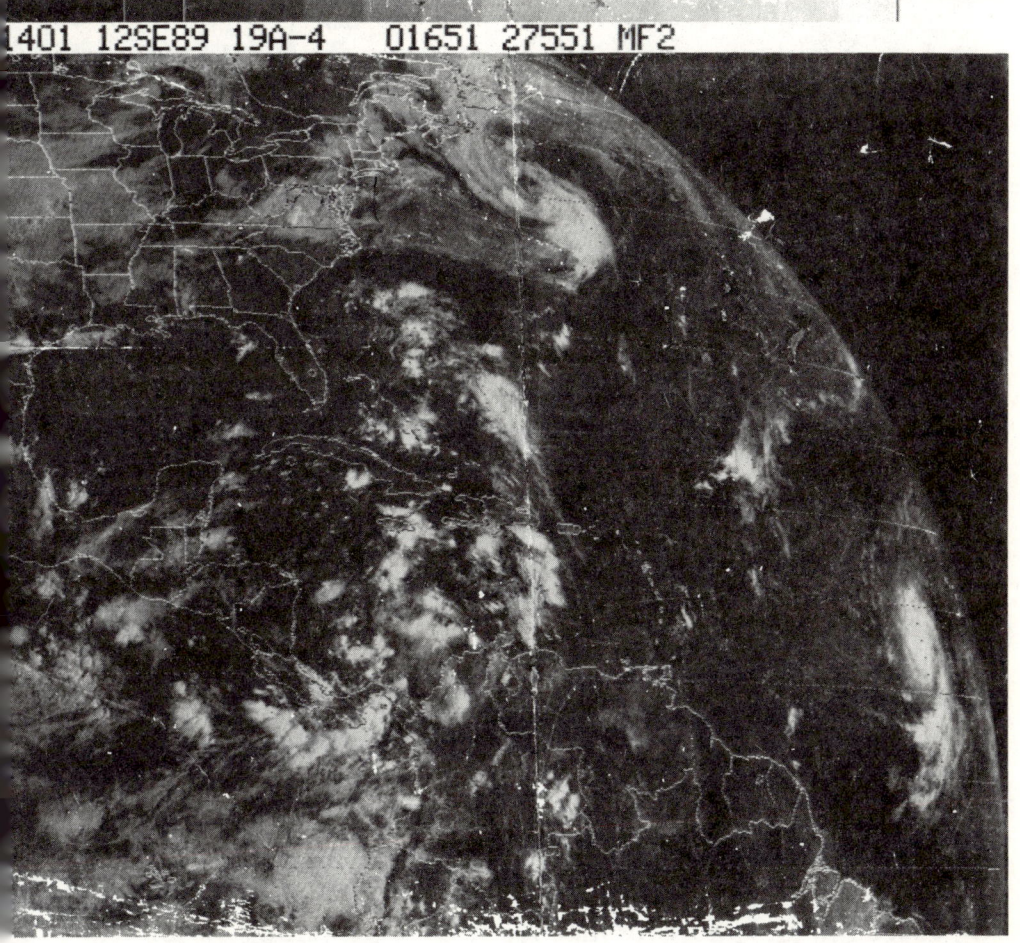

Satellite picture of Cyclone Hugo, 300 nautical miles to the east of my course

9: PARAMARIBO TO NATAL

I soon found out why people advised me to avoid Paramaribo. Even as I taxied to my parking bay I could see that the airport buildings reflected long neglect. Paint seemed to be something that disappeared with the Dutch colonial administration and every building showed signs of decay. It had an aura of 'days of glory passed'. Jungle surrounded the airport and, in many places, seemed to be biding its time, waiting to claim back what had been taken from it.

There were a large number of soldiers around and they swaggered in an aggressive manner as though waiting for an opportunity to use their rifles, and one of them, in particular, threw his around in a very threatening way. I couldn't help thinking of my landing at Johnston Island and the precise and disciplined way the military there had handled my intrusion. This was different. These soldiers seemed to have no discipline and it was for that reason they were so intimidating. I could tell I was not going to be offered any ice cream here.

As I got out of my plane a civilian approached me from the tower and introduced himself. His name was William Van Biene and he was the controller who had arranged permission for me to land. He was also the person who had handled the original application which I had made from Australia to land at Zendary, so he knew of me. This was a lucky break.

William helped to expedite my paperwork, customs clearance, and so on. As at all airports where I landed I organised fuel for my onward journey. Knowing that I had enough fuel for the next leg allowed me to sleep easier at night and also saved a lot of time in the mornings. In some places there were hassles because of communication difficulties and if I did not know the procedure at that particular place.

After organising the fuel, I would go through customs, and immigration, and health checks, pay various fees for landing etc, then I would think of a place to stay. Usually I'd find help in the briefing office, where I would also call to ask for the procedures for putting in my flight plan for departure, and enquire about weather information. I found it important to be diplomatic as the briefing office personnel could make it all easy, or impossible. Mostly, they were wonderful, especially when they knew what I was attempting to do. Often they marvelled at a woman alone, flying around the world. Frequently they'd ask, 'Where is the rest of your party?' They'd stare in disbelief when I explained that I was the only one. Sometimes they looked like they were about to faint. Then I would ask for help with taxis and hotels.

This time William found me a cab driver who would take me on the hour and a half drive to the town. He introduced me to Twist, one of the few cab drivers in the country who sort of spoke English. William said that Twist was a friend of his and I could trust him. Twist and I agreed on a price, which William told me was fair. For this Twist would take me to the best hotel in town and pick me up the next morning to take me back to my plane. I made sure that the plane was secure for the night, because, despite all the military personnel around, I really didn't trust anybody.

I was appalled by the poverty and ruin I saw as we neared the town centre. The State buildings, which once had been very grand, had seen no maintenance since the Dutch left, and the hovels and poverty of the people an indictment on the people who ruled Surinam, but certainly didn't lead it; it looked to me like a country going nowhere. I was also struck by the number of very ill, mangy dogs around the streets and thought about the pampered pets back home. I imagined the reactions from organisations such as Animal Welfare. But I could see that these poor people, who spent most of their lives scared and wondering where their next subsistence meal was coming from, hardly had the time and the resources to look after themselves, let alone the dogs.

My 'international' hotel was no surprise: threadbare carpet and a profusion of smells that included mould and insecticide. Frankly, it was horrible. I tried to phone Australia, but soon found that I couldn't make a phone call out of the country, although I could send a fax. I faxed my sponsors.

That night I walked into a bar in the hotel to get a drink and found myself the only white person in a group of over one hundred. I was an absolute minority and decided not to stay. The people weren't unfriendly, and in fact were courteous, but somehow I felt uncomfortable and wondered at the difference between Surinam and Barbados, where I felt that I fitted in well. Barbados was not a wealthy country, except in the generosity of its people, who were cheerful, with a pride and sense of worth that was lacking in Surinam.

But it was all right for me to compare and be offended by the poverty of the Surinamese and then puzzle at their condition. I was flying out the next morning. They had no such escape.

After Twist drove me to a Chinese restaurant for a meal, and brought me back, I slept very soundly and woke at 6.30 am. My room smelt strongly of damp so I quickly showered and hurried downstairs for a light meal. At 7.30, just as I was finishing breakfast, I was summoned to the hotel desk and told that there was a phone call from Australia for me. It was my sponsor, who had received my fax and decided to see what could be done to cheer me up.

Twist arrived to pick me up on time, with his lovely, warm smile, gold teeth and earing. We drove through the dilapidated streets and I couldn't help thinking how hard it is to crush the human spirit.

There was the usual delay at the airport as I filled out the necessary paperwork. I think I must have filled out several reams of paper on my way around the world. Every country has its bureaucracy and they all demand satisfaction. But I eventually took off for Cayenne, overjoyed to be out of Surinam. The cloud was already beginning to build up, but it was a short flight of only one and a half hours, so I wasn't greatly concerned.

The contrast between French Guiana and Surinam was startling. The French controllers were pleasant and aloof. From the moment I landed I spoke French. I had almost mastered the language during a stay in France many years before and I appreciated the fact that I could communicate once more in this demonstrative language. I was a little apprehensive at first, but it all came back. I was amazed at myself.

I hadn't organised a visa, as Cayenne was not on my original flight plan, but the airport policeman who was handling my paperwork kindly gave me a transit visa. How easy it all was. *Tout va bien encore!* I gave out some of my Aussie stick pins — koalas and kangaroos with a chip of opal. They were well received.

I caught a cab to town and the driver recommended a good hotel: *'C'est un bon hotel, qui s'appel Hotel Amazonia.'* Thank God for the Hotel Amazonia. Cayenne was clean and well ordered and I felt safe walking around the streets. The population was mixed, with a lot more white people than in Surinam. The cars were Peugeots and Citroens, and all in good condition.

I asked directions to a good restaurant and found myself in *Le Jardinier.* It was like a rural restaurant in France, with a tantalising cuisine and a typically jovial French clientele, except perhaps for the skin colours of the people and the beautiful, brown-skinned children. As I sipped my bordeaux and munched on the exquisite fricasse of tabby cat? I mused on the variety and incongruity of life. *Quel bonus!* To find this in the jungle.

My phone call to Boston Bob for a weather report was not very promising. He advised me not to fly the next day as the unsettled weather conditions thrown up by Hurricane Hugo were still affecting the weather and throwing big storms on my path to Belem in Brazil. But I didn't mind the enforced stay. My hotel was wonderful, and reminded me of the Hotel Roma in Montmartre, where I had once stayed when I was very young.

The next morning's breakfast produced a pleasant surprise which dispelled any concerns about what to do with myself. I got into a conversation with a school teacher who had recently transferred from France. His name was Jean-Pierre Louis and he told me that he was waiting for his wife to arrive from Paris. He offered to show me the sights, an offer which I readily accepted. I don't think he quite believed that I was actually flying around the world, and when I told him I would first have to go to the airport to check my plane, he quickly agreed. It was fun to see the look on his face when he saw my plane, which looked tiny against the Air France jets he was used to.

That night Bob gave me the all clear for the next day. I decided to get away early to avoid the cloud build up over the jungle. It was only a 4 hour flight, but one I was not particularly looking forward to. Radio communication was almost non-existent, although I was getting used to that.

I was on my way at 8.00 am and soon found myself across the border into Brazilian airspace and flying over the jungle. Wherever I looked I saw the jungle and my old fears soon returned. What happens if I go down in this? I'd rather be flying over the water! No, no. Hang on, there are sharks in the water! Sharks!! At least I've got a vest and life raft. What survival gear have I got for the jungle? None! I could see the headlines: 'Aussie flier lost in Amazon Jungle'.

Give over, Gaby. What are you doing, creating a disaster when none is likely? Just keep your eyes on the instruments and make sure you're on course and stop this stupid worrying!

I've got to admit that there were times that my imagination ran away with me and I imagined the worst. In fact, the times I had come close to misfortune had happened when I was feeling confident. There was little point in imagining myself into some tragedy, so I settled into the flight and began to enjoy it.

The Amazon jungle is unique. Flying over it gave me an appreciation of its immense size. It seems to just go on and on in a continuous sea of green that contributes to the oxygen we breathe. I found it hard to imagine what the world would be like if all of this no longer existed.

The maps told me I was flying over true wilderness. 'This area unknown' and 'elevation unknown' they declared, telling me I was over unexplored terrain. One of my sponsors, QANTAS, the Australian international airline, had given me considerable logistical support before I left Australia. They helped me plan my route as well as organising landing clearances and overflight clearances. I recalled their great concern that I would be flying over these areas in South America. There was no radio contact and I was, virtually, on my own.

I flew over part of the Amazon delta, which is at least 60 kilometres wide. The jungle would give way to mudflats, then water, then mudflats, then jungle again. It went on and on and half an hour passed before I cleared the last stream. The water was very brown and continued for hundreds of kilometres into the Atlantic Ocean. I wondered how long it would take to wash all the top soil into the ocean if the jungle were destroyed.

I established radio contact with Belem when I was 50 miles out. It was good to talk to someone again, although I had doubts that we were communicating. English is the international language of the air and all controllers are supposed to conduct their conversations in English, but, as I neared Bellem and came under their control, I realised that the controllers and I had learnt English in different schools. I was unable to understand them.

I asked several times for clarification and landing instructions and was greeted with silence. It was frustrating and, in the end, I took matters into my own hands and declared that I intended to use runway zero-six. There was no reply, so I put my plane down on runway zero-six. Fortunately, I could make a visual approach and it was a good landing, which was the last good thing to happen to me that day.

I was escorted from my plane to the offices of the airport authority and asked for my papers by a federal policeman. I passed over the normal papers relating to my flight, but that wasn't enough. I was told to get every paper relating to my plane and my journey. This meant trudging back to the plane and collecting the indexed file which held just about every piece of paper I owned. I felt frustrated and angry as I trudged back to the office, but I knew what I was up against. *You'll never beat the bureaucrats, Gaby.*

I have dealt with bureaucracies in many countries, but none, in my experience, matches the love of the Brazilian bureaucrat for paper and the apparent importance that goes with it. Every paper was minutely examined and the airport authority found I didn't have the original plane registration. I talked my way out of that one, by proving that all the modifications to the plane were in my name, and thought I was going to get the okay, when the federal policeman discovered my visa was past its expiry date, because, of course, I was running very late on my planned schedule.

Oh, oh. We're in trouble, Gaby. We were up the proverbial creek without a paddle. I think I should have offered him money, in retrospect. I believe that would have settled the matter, but there are times when I'm none too fast on the uptake and that was one of them. I guess I can plead ignorance, becuase that way of getting things done is not part of the Australian way. I offered the policeman a souvenir of Australia, a small kangaroo with an opal chip. He demanded another one and adopted a belligerent manner.

'Pig,' I thought. I was about to tell him what he could do with his rude manners and his country, but I bit off the words. I was sure he could read my expression, though. Okay, we'll stick this one out. A television crew from Channel 9, based in California, was due to fly in. They had a Brazilian interpreter. I thought I could hold out until then.

At 5.00 pm he eventually told me I could stay in the country for three days. I took the paper he offered and virtually ran out of the office, hoping the television crew could solve my dilemma, as I knew my plane needed a major service and I would need more than three days in Brazil.

I caught a cab to my hotel, the Equatorial and, though it was old-fashioned, it was comfortable and clean and, more important for me at that time, the staff were pleasant. Only two things marred my stay there: the traffic noise (it was on a major road), and the cockroaches, which were the biggest I had ever seen. Someone once told me that some of the biggest 'roaches in the world are in the Australian forests, where they eat the vegetable matter on the forest floor and are part of the cycle that puts nutrients back into the forest. They are strictly forest dwellers and shun mankind. I've never seen them, however, and doubt they could be as big as the domesticated monsters that inhabit the world's equatorial regions. I'd put those Brazilian 'roaches up against any.

Belem is a city with a population in the millions and is regarded by most Brazilians as a backwards frontier town. It's on the north-east coast near the mouth of the Rio Tocantins, away from the sophisticated southern population centres like Rio de Janiero and Sao Paulo and is the only major population centre before jungle wilderness. It really is an outpost, isolated from the rest of this huge country. The old city shows its colonial past and is peppered with reminders of the colonial ambitions of the European world. The Portuguese were the dominant colonial power and theirs is the language spoken by the population, which is a rich mixture of southern European, African black and Indians.

The Channel 9 crew showed up at the hotel about 8.00

pm. Producer Julie Ladner had also handled the filming on the United States mainland. With her were Hal Landen and Merce Williams, freelance cameraman and sound recordist, and Christiana Mesquita, our interpreter from Rio. I was pleased to meet them and we quickly decided on dinner in the old town. The place chosen was an old fort on the river. We walked on cobblestones between massively thick walls to the restaurant, which was in the old barracks building. The setting was exotic.

We ordered a fish meal and I recounted my problems, which seemed to disappear as the beautifully prepared and presented meal arrived. I looked around the restaurant and saw all the Portuguese-speaking Brazilians enjoying themselves. How thankful I was to have Christiana with us. She and her husband worked as a film production team, based in Rio. She is also a linguist and fortunately Julie had known of her. I drew a lot of comfort from Christiana's local knowledge and supportive character.

Australia doesn't have a diplomatic office in Belem, although there is an embassy in the capital, Brasilia, many kilometres away. The next morning we went to the Honorary British Consul, Mr Robin Burnett, a true English gentleman, who had lived in Belem for thirty years and had a vast knowledge of Brazil and the wonderful ways of its bureaucrats. Fortunately, the local Federal Police Chief was a good friend of his, so Robin arranged for us to see him.

There are times when being accompanied by a television crew is an advantage, and times when it is definitely not. Julie thought it would be a good idea to film me entering police headquarters. Christiana didn't think it would be such a good idea and, after my experience of the police the day before at the airport, I *knew* it wouldn't be a good idea. However, Julie would not be swayed. It would make good atmosphere, she explained.

It certainly didn't make good atmosphere with the police, who are paranoid about their image — expecially as it is perceived in the West. As I walked across the street and entered the front doors of the police station with the television crew set up on the footpath opposite and filming away, the police trundled out the door and quickly grabbed Hal and Merce and hustled them into the building and down some stairs and into a cell. We had definitely blown it. Christiana had to do some fast talking to get them and their gear out of there. The Chief of Police refused to see me. It was fast becoming a farce.

We went back to the hotel to decide on our next move and, fortunately, Her Majesty's Consul phoned us there to find out what we had done to upset his friend. Once we explained what had happened Robin undertook to try again and did arrange a meeting with the Chief of Police for later that afternoon. As I had to arrange a service for my plane, we went to the airport and the farce continued.

Julie thought the atmosphere could be improved if they filmed me speaking by phone to the Australian Ambassador in Brasilia. Now, the Brazilian economy suffers from an inflation rate that soars with the speed of a ballistic missile. The military government at that time overcame this in a very simple way. They devalued the currency every few months or so to make allowance for this. Of course, the telephones at the airport are coin operated and to make a long-distance call required a large number of coins, which had to be fed into the machine at an incredibly fast rate to avoid an automatic cut-off. I'm afraid my reflexes weren't up to it and halfway through my call to the Ambassador in Brasilia I missed the coin feed and the call was disconnected.

I was feeling pretty angry by this time and I stalked out of the airport office to inspect my plane and to arrange its service. As I walked across the apron with the Channel 9 crew following I saw two men looking at my plane. They were two pilots employed by a large Brazilian bank and had just landed in their company's executive jet. They quickly introduced themselves and, after hearing of my difficulties, gave me some invaluable advice on how to make my way around the bureaucracy.

I finally got my plane into the hands of the service company Lida Aviation and their chief mechanic, who went over the plane with me. I had noticed a small oil leak on the way down from Cayenne and there was hydraulic oil on the floor of the cabin. I also explained about the problems I had experienced with the auxiliary tank switching system and he told me that he could probably find the problem and fix the switch and attend to the oil leak as well as give the major service. I was very relieved, even though all of this would probably take three days and I didn't have permission to stay for that amount of time. I didn't want to have to leave Brazil before the plane's service was finished.

The interview with the Chief of Police was difficult for me. He reminded me of one of the nuns from my early school days. He was very stern and did his best to make me feel guilty for something I had not done. After a while he calmed down and could see that I was obviously not a threat to the internal security of Brazil and that the television crew were equally harmless. Julie had mentioned the possibility of going on a boat ride the next day, so I thought I would ask him if this was okay. As soon as I asked his permission for something his grim mask dissolved and the doors opened. He became very talkative and wanted to know all my plans.

I told him I was planning to fly to Fortaleza and then to Recife, where I would take off for my Atlantic crossing to the African coast. I explained how Amelia Earhart had taken off from the airforce base at Natal, but I couldn't arrange clearance for there. He immediately got on the phone and arranged for me to leave from Natal — amazing! And when I walked out of his office I had a permit to stay in Brazil for ten days.

I could relax now. I phoned my children and caught up on my correspondence, making sure that Jamie's class would receive a good supply of postcards in the mail. Before dark I went for a short stroll around the streets near the hotel. I saw a lot of poverty and many apparently homeless children. The greatest shock came when I turned a corner and almost trod on the torso of a woman in front of a begging bowl on the pavement. She was, virtually, *only* a torso. I became indignant and self-righteous. How could anyone exploit a person's disability in this way? I fumed inside.

Hang on, Gaby, it isn't like that. Look around at the poverty around here, Gaby. Perhaps she is the breadwinner for her family. Perhaps for them her birth was a blessing and she is being of real service to them. Imagine, in some civilised countries she'd be locked away, but here, where there is more tolerance, she is out on the streets with people. Look again.

I had always thought of myself as tolerant. Perhaps the journey was testing my tolerance. On the streets of Belem I found I was not as accepting as I had imagined myself to be. I was discovering depths of intolerance that I didn't know I had. While the beggar woman's beautiful eyes smiled at me, I found my own eyes filling with tears, but they weren't tears of compassion, they were more like tears of frustration. I turned away and walked back to the hotel in a reflective mood.

The next morning we went down the river and climbed aboard the Amazonian equivalent of the *African Queen* for a cruise on the backwaters of the Tocantina River. We cruised along the tributaries and creeks, where the jungle canopy met over our heads and the wildlife, although mostly unseen, could be heard. Occasionally we'd pass some hut built on stilts over the muddy river bank, or some local people would glide by in a canoe. I was in a different mood.

At lunchtime we pulled into a thatch-roofed hut on the riverbank. It was isolated and surrounded by jungle. It stood on stilts, with a small jetty which reached out to deep water. Inside there was a small dining room where we were served Cokes and beers out of an old kerosene refrigerator. There was a kitchen out the back where a young woman was preparing our meal. She had a very young baby with her and I picked it up and started talking to the mother. We had a great conversation, me chattering in English, she in Portuguese. We couldn't understand a word the other was saying, but we were two women sharing a part of our life together.

Our meal consisted of chicken and a Brazilian speciality
— an animal called a Bacau, which looked to me like a cross
between a pig and a dog. The flesh was like pork, and
fortunately I ate little of it. I found out later that the crew
came down with food poisoning after leaving Brazil. The
kitchen was clean but very primitive and the scraps went
straight out the window to the mudflats below where a dozen
or so scrawny hens were busy recycling the household
garbage.

After lunch we made our way back to Belem and stopped
at a pavement cafe for a coffee. We were approached by a
number of young boys, about six or seven years of age, who
were selling nuts wrapped in used computer print outs.
Christiana told me that there were about 15 million homeless
kids in Brazil — almost the population of Australia. On the
way back to the hotel I saw a young boy asleep in a doorway,
with no covering, no shirt and no shoes, but somehow
surviving. I though of my own beloved seven-year-old son.

My plane was now ready and Lida Aviation had done
a good job. They'd discovered that the hydraulic oil reservoir
was empty. This was a potentially disastrous problem.

I was happy to get out of Belem and on with my journey,
so the next morning I was out at the airport early to get
away to my next stop, Fortaleza, further south on the coast
of Brazil.

I went to the airport office and paid all my fees. It took
me an hour to complete the formalities and, as I taxied my
plane to the runway, thinking that I had escaped the clutches
of Belem's bureaucracy, I got a call from the tower telling
me to report back to the office as they had discovered
another fee. I cursed at the delay, thinking of the cloud build
up. *Oh well, you can't beat 'em, Gab!*

I finally got away and took off over the jungle. As I neared
Fortaleza the countryside changed dramatically, from steamy
jungle to dry and sandy coastal plain. There was a strong
nor'easter blowing off the Atlantic.

I was looking forward to my short stay at Fortaleza as
Christiana had told me about the wonderful king prawns that

are caught in the waters off the coast. My landing wasn't the best because of a sudden wind gust, and also because I had been ill — presumably the results of the Bacau — and I was still feeling queasy. After a press conference I went to my hotel and went to bed. I missed my dinner of prawns. That would teach me to think about food.

I felt better by breakfast and, as my next leg was a short hop to Natal, I took my time and got away at 10.00 am. On the flight I noticed hydraulic oil on the cabin floor again and became concerned. I knew I couldn't possibly attempt the Atlantic crossing with such a major problem. Somehow or other I would have to get it looked at. My only hope seemed to be the Brazilian Air Force at Natal.

There are many things about Brazil that I didn't like, and the bureaucracy was at the top of the list, but balancing my list of dislikes was a much longer and more important one made up of my very real likes. Heading that list was the people and none was more generous or helpful than the Commanding Officer, and his men, of the Brazilian Air Force at Natal. When he heard of my plight he organised for his engineers to check my plane and repair the leak. It was a big job as the hydraulic reservoir was tucked away underneath my first auxiliary tank in the nose luggage compartment, and the tank had to come out first before they could even look at it. It meant a 24 hour delay, but it also meant safety. I was pleased to sacrifice this time, especially when I phoned Boston Bob and heard his weather report. I could expect severe electrical storms across the Equator.

My hotel was right on the beach and I was surprised to see the topless and almost bottomless costumes which displayed so much flesh. It reminded me of the beaches around Sydney, particularly Balmoral. The next morning I walked along a deserted beach on my own. It was good to put my toe in the Atlantic. This was the water I was going to cross that night. As I walked on the sand in my reverie, I had the sense that I was walking on Palm Beach, where Neville and I had often walked together.

Before checking out of the hotel I phoned Bob for an

update on the weather and he told me I could again expect thunderstorms about 5 degrees south of the Equator. I would be flying through the Inter Tropic Convergence Zone (ITCZ) where the southern and northern air masses meet and at this time of the year it is very turbulent. It is in this zone, off the coast of Africa, that the great storms and hurricanes are spawned. It was something I was not looking forward to, yet I knew I had to face it.

My departure was planned for 5.00 pm, which would give me the night to fly the 1600 miles across the Atlantic and arrive at Dakar, Senegal, at first light.

I arrived at the Air Force Base with plenty of time to spare and the chief engineer and his men explained that they had checked over my plane and repaired a small leak in the hydraulic system. If left unrepaired the oil would again have drained away and the hydraulics, that is the landing gear, would have ceased to function. I was endlessly grateful to them.

I was given a tour of the base and, at 5.00 pm, I was in my plane and taxiing to the runway for take-off. The plane was heavy with a full load of fuel and my gear. I was pleased it was such a long runway.

As my plane gathered speed I saw five jets in formation before me. This was the Brazilian Air Force saluting me and giving me a fighter escort out over the Atlantic. My wheels lifted off the ground and I began the slow climb to my cruising altitude. The fighter pilots' voices came over the radio, wishing me well as they peeled away. I thanked them and silently said a prayer wishing myself well. I wondered if the Atlantic would prove to be as dangerous as the Pacific.

*Amelia Earhart on the steps of the Civil Airstrip Building, Fannie Bay,
Darwin, June 1937.* COURTESY R. N. ALFORD

*Amelia Earhart, Fred
Noonan and an
unidentified official at
Fannie Bay, June 1937,
walking from her Lockheed
Electra 10E, NR16020.*
COURTESY R. N. ALFORD

▲▶
*Amelia prior to her
departure from Lae, Papua
New Guinea.*
COURTESY MRS K. FRANKLIN, TAKEN
BY GILBERT SINCLAIR FRANKLIN,
FROM BULOLO GOLD DREDGING
COMPANY.

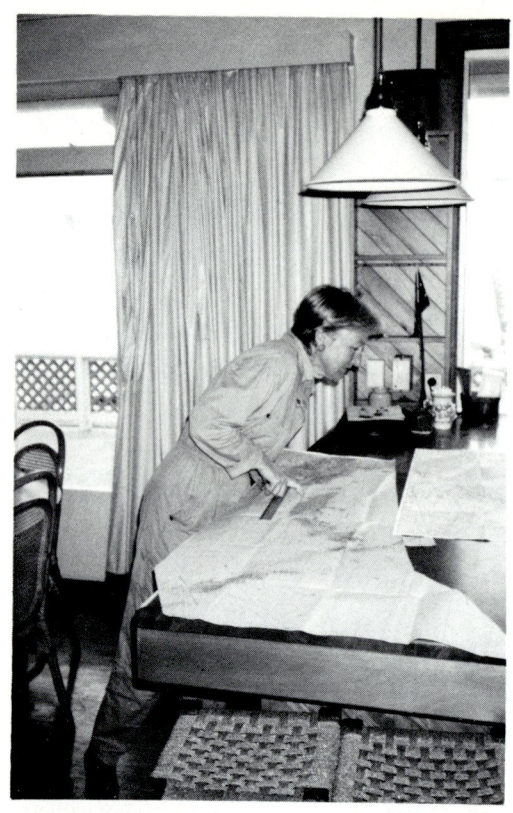

▲

Shortly before I departed, and part of the lengthy planning.
COURTESY LYNORE BROOKE

▼

Last minute preparation of the auxiliary fuel system. I'm there to help by doing odd jobs.
COURTESY SHERI FORBES

Dick Smith, on behalf of the Civil Aviation Authority, presents me with a special pilot's document case on 31 July 1989, a few days before my departure.

Departure day, 3 August 1989, and it's time to say goodbye to my children — we're all being brave!

Just prior to departure, 3 August 1989, in my crowded cockpit, surrounded by fuel tanks, my maps and flight charts beside me.

My crowded cockpit from the back door, showing fuel tanks and luggage.

▲ *Departing Bankstown, 3 August 1989.*

▶ *August 1989, Lae, Papua New Guinea. The Papua New Guinea Air Force commanding officer and visiting RAAF officers pose after Amanda Wright, Ella Birrel and I had laid a wreath in honour of Amelia Earhart and Fred Noonan.*

▶ *Lae, Papua New Guinea, and the strip Amelia Earhart last took off from.*

The commanding officer took this picture of me as I was about to leave Johnston Island. It looks very touristy — what a false impression!

Dave Dennis, the Mayor of Atchison, Kansas, presents me with two keys to the city on my arrival, and makes a proclamation of Gaby Kennard Day, 5 September 1989.

COURTESY DONALD E. MARTIN

*Motorcade through the
streets of Atchison to the
Forest of Friendship, with
Virgene Smolik driving.*
COURTESY DONALD E. MARTIN.

*The 99s and the people of
Atchison commemorate my
flight with this plaque in
my honour. I was hoping,
for their sakes, as well as
for my own, that I would
finish my journey.*
COURTESY DONALD E. MARTIN

*In the Forest of Friendship,
Atchison, Kansas, 5
September 1989. Side by
side with Amelia.*
COURTESY DONALD E. MARTIN

*Checking the oil, Memphis,
Tennessee, 9 August 1989.*
COURTESY NANCY MILLER

Amelia Earhart's birthplace, Atchison, Kansas.

The Indian Women Pilots present me with a silver plate commemorating my solo around the world flight, in Bombay.

On Australian soil again. 4 November 1989, and I sign autographs for children in Darwin. The people of Darwin gave me a wonderful welcome.
COURTESY NORTHERN TERRITORY NEWS

Nearly home. Taxiing in for a briefing with the Royal Aero Club formation fliers, Camden, 10 November, 1989.

140

*The Royal Aero Club
formation fliers, with me
over Bankstown Airport, 10
November 1989.*
COURTESY BURNIE PHILLIPS

*About to touch down on
runway 29er right,
Bankstown, 10 November
1989.*
COURTESY BURNIE PHILLIPS

*Taxiing in at Bankstown,
10 November 1989. I'm
home and all those people
are waiting to greet me in
the rain.*
COURTESY BURNIE PHILLIPS

Hugging James and walking through the crowd at Bankstown, 10 November 1989. How wonderful to be home.
COURTESY BURNIE PHILLIPS

Starting on the motorcade through Sydney, with Mimi and James, 10 November 1989.
COURTESY BOB LIVINGSTONE

Excitedly arriving at Sydney Town Hall, surrounded by well-wishers. This was such an emotional time. These people are really here for me. How amazing!
COURTESY BOB LIVINGSTONE

Those wonderful members of the Australian Women Pilots, who helped and supported me all the way, at the Mayoral reception, 10 November 1989. On my left is Nancy-Bird Walton.

▲
Together again with my children.
COURTESY TELEGRAPH

10: NATAL TO DAKAR

I was alone again.

The 1660 nautical miles began to stretch before me as my plane pulled its heavy load of fuel on the long, slow climb to my initial cruising altitude of 5000 feet. Perhaps I didn't really need to carry a full load of fuel, but my encounter in the Pacific prompted me to err on the side of caution and I felt happier with the safety margin the extra fuel gave me. This time I've got a properly functioning auxiliary tank selector, I congratulated myself, as I looked down at the clear display of the shiny brass handle and new tank numbers. The people at Lida Aviation had done a great job and it looked very professional — and very reassuring.

My perspective of earth broadened as I climbed higher, and I saw again the semi-desert that surrounds Natal. I had first noticed on my journey from Fortaleza the stark contrast of the country around Natal to the lush jungle further north. It prompted my curiosity enough to enquire of a local. The answer was simple — sugar cane had been intensively grown here more than a century ago and in 150 years the land had degenerated from verdant jungle to desert. I'm sure mankind can learn to live harmoniously with nature, but every country has examples of this sort of exploitation. There are many examples of this in Australia. We humans don't seem to learn.

Short term profit so often wins out at the expense of the long term. But at least the awareness is growing, and in that there is hope.

Night fell as I flew over a tiny island called Noronha 150 nautical miles out from the coast. There was a small airstrip on the island and I could see the lights of the houses in the quickening dark. I called the control tower and reported my position. Whoever was on duty acknowledged my call in Portuguese. Well, I thought, we may not be speaking the same language but at least we have communicated. There are times the sound of a human voice is very comforting, particularly on a moonless, pitch black night 5000 feet above the South Atlantic.

I was supposed to be in contact with Recife Centre, as I was in Brazilian air space. However each time I tried to raise them I received no acknowledgment. I was warned that this might happen as it seems that at night Recife is manned by juniors who do not speak English. I doubt many people have flown this lonely ocean at night since Amelia Earhart did in 1937.

I couldn't care if Recife Centre spoke gibberish, all I wanted was to know that I wasn't alone on a pitch black night and that someone was aware of my presence. All I needed to hear was the sound of a voice. But I eventually gave up on Recife and decided I would try and raise Tony Vacarella in Sydney, but this didn't work. I tried on different frequencies and just as I was about to give up, I heard a voice acknowledging me in English. I hadn't raised Sydney, but through some miracle I had managed to make contact with Portishead Radio, a British Telecom world-wide aeronautical and maritime radio service operated from Somerset, England. What a relief.

When Julian and Will at Portishead Radio got over their surprise at being raised by an Australian woman solo pilot flying across the South Atlantic in a single-engined light plane, they came to my assistance and made contact with Tony in Sydney, who gave them a frequency for me. However I still couldn't raise Tony, and when I reported this to Julian he promised to make contact with me every 30 minutes.

I also asked them if they could give me the frequencies of any broadcast radio stations in Dakar, thinking I could use one of these as a homing beacon, similar to an NDB. Julian came back to me with the news that he couldn't locate the frequencies. But then Peter Cox, a ham radio operator in England who had picked up our conversation, passed the frequencies to Julian for me. How great it was, I thought. From being without any contacts I was now in touch with Julian and Will and ham operators could hear me.

I tried the frequencies Peter Cox supplied, but unfortunately the Dakar radio stations weren't broadcasting 24 hours and had shut up shop for the night. Oh well, it was a nice try.

I was now over 1000 nautical miles out of Brazil and had made my third crossing of the equator. I was eight degrees north and well into the Dakar Ocean Flight Information Region, which meant I should be talking to them and would have to sign off from Julian and Will. I thanked them profusely and hoped I had conveyed the appreciation I felt for the comfort they gave me.

According to my calculations I was through the position Bob had given me for the thunderstorms and was beginning to think I had escaped them, when I saw ahead of me the faintest flash of light — the harbinger of the wildest flight of my life. I couldn't have avoided those storms even if I had weather radar installed. It was an option I'd looked at in Sydney but the cost had put it out of my budget. I would certainly install it if I was going to undertake a similar journey.

The lightning flashes started coming more frequently and all of a sudden I was in the middle of it. Massive flashes lit my cabin and my plane was beginning to pitch and toss all over the sky. The lightning would flash to my right and I'd head thirty degrees left to avoid it. Then it flashed in front. I altered forty degrees right. It flashed left. I veered right. It was like 'The Charge of The Light Brigade': 'Guns to the right of them, guns to the left of them, guns to the

front of them volleyed and thundered.' I was suddenly flying
for my life.

I wanted to use the porta potty but didn't dare let go
of the controls. My plane was like a cork in a wild sea. I
would hit a downdraft and within seconds be sucked down
2000 feet. The next minute a swirling updraft would carry
me up 3000 feet. Anything loose in the cabin of the plane
began flying around, and I became worried that something
solid might hit me on the head. I had never experienced
anything like this.

The storm cells were lined up, row upon row of them,
right in my flight path. I'd get through one and have a minute's
breathing space when the next one would rush upon me.

I was in the thick of them when my electrical system
started playing havoc. The ammeter suddenly showed a 35
volt drain as if I'd turned something on. Then my gear unsafe
warning came on, which indicated that my landing gear was
in transition and this was proved true within seconds when
my forward speed dropped from 130 to 90 knots indicated.
Then, as the gear retracted, speed increased again and the
light went off. This kept happening and I couldn't work out
why. I had the emergency gear override system on, which
I think mercifully kept the gear from fully extending, but it
was definitely coming almost all the way out, as my indicated
speed kept fluctuating.

My green transponder light, which indicates when a
radar station has me on screen, started to flash in a bizarre
way and I noticed the Omega lose power and go off the
air momentarily. It came back on and tried to re-position
itself. I turned off everything except the Omega, which
appeared to be totally confused. It didn't work again. It had
lasted from Memphis to now. I was grateful for that at least.

I endured this turmoil of varying degrees of severity for
three hours. I had no other recourse but to go through it
and, eventually, I did. As suddenly as they came upon me,
the storm cells had gone, leaving me in a state of exhaustion
and not fully able to comprehend what I had been through.
These Atlantic thunder storms range in height from 1000

to 50,000 feet and each one carries more energy than an atomic bomb. I wouldn't have dared to fly under them as the up and down draughts are severe, and at least with altitude there is room for vertical movement. Somehow I had survived, thanks to my sturdy aircraft and the engine, which kept turning without missing a beat.

I had wanted to let someone know what was happening to me, because at one stage I didn't think I would make it. I tried to explain to Dakar control, but they didn't want to know. In retrospect I think my English was the problem — they didn't speak it very well and kept asking me, 'What is your estimate for your next position?'

Has it gone, Gab? Oh, so you're back. A fat lot of good you were. *I was scared.* Well, so was I, but I couldn't go off somewhere and hide. I had to stay here and get us through. *I know. You did a great job, Gab.* I don't know about that. I think we've got God looking after us on this trip. *I hope so, Gab, but God couldn't have done it without your help.* I *was* pleased. This was another test I had gotten through — somehow or other. My skills as a pilot came to the fore as an automatic response to the emergency, but it was more than skill that got me through that night, I really had the sense of being looked after.

I was approaching Dakar, about 100 nautical miles out, when my Distance Measuring Equipment (DME) came in momentarily. It was enough to tell me how far from Dakar I was. It came back on at 70 miles and held. I was overjoyed as my VOR indicated I was exactly on track. I had planned to arrive at Dakar in the early morning, but a tail wind component had pushed me along and I had made better time than expected. It was 6.15 am and still pitch black. I slowed down.

The tower gave me an Instrument Landing (ILS) approach, which I handled in a very ungainly way, but I touched down on the west coast of Africa as dawn broke. I was more than halfway on my journey.

When Amelia Earhart landed at Dakar it was the capital of the French colony of Senegal and she was feted by the

French officials. Senegal is no longer a French colony, although the French influence still lingers a little. Unfortunately, it is now the most malodorous place in the world: dirty, untidy and very smelly. The smells began at the airport and stayed with me long after I left.

I was met at the parking bay by a very personable and effusive Senegalese called Mamadou. He was from a local charter service and took me under his care in a very gentlemanly way. I told him about my problems with the electrics and he directed me to park at the Aero Club and helped me organise someone to check out the plane's electrical and hydraulic systems. He then helped me through police and customs and directed me to the La Cayou hotel, which was set out as a series of thatched huts on the beach. It was run by French-speaking Lebanese and although it boasted 'International Accommodation', it fell somewhat short of that mark, but was a lot better than I had expected.

I checked in and walked to my hut, opened the door and almost gagged from the smell that came out of the bathroom. The smells that issued out of that room were too horrible to describe. Everything looked reasonably clean but somehow or other the smell crept through from God alone knows where. There's a saying in Australia that describes something as smelling worse than a country dunny. This was worse than that.

I shut the bathroom door and turned on the air conditioning and this seemed to relieve the assault on my olfactory system. I went to bed exhausted from my ordeal over the Atlantic.

It was late afternoon when I woke and I quickly showered and dressed and took a quiet stroll along the beach. It was strange for me to see the sun set over water. I was used to the beautiful sunrises that climb out of the Pacific and grace the eastern coast of Australia.

The hotel dining room was an unusual arrangement and looked like a long hut set amongst the smaller accommodation huts. I noticed as I was being served that the waiter didn't have very clean hands. I wish that I had paid more attention to that warning. I finally got to taste the Atlantic king prawns and they were a real treat. I also had some spicy dish wrapped in lettuce leaf and heaped with spring onions.

I met a group of French-Canadian and American agricultural pilots who were staying at the hotel. They were under contract to spray locusts and had just finished a tour in the backblocks of the country. They told me some horrifying tales of life in the interior. I told them of my flight plans and how I was going to divert from Amelia Earhart's flight path at Dakar. Amelia went due east across the African continent to Assab. It's amazing that what was possible in her day can no longer be done because of the number of wars now taking place in the countries along her path. To avoid this I planned to fly up the West African coast to Morocco and then north-east to the Mediterranean across to Crete and south through Egypt to Luxor and then east across Saudi Arabia to Bahrain. My flight path represented a major diversion from Amelia's, but I did intend to arrive home in one piece, and being the target of a surface to air missile or .50 calibre machine gun, was not the way to do it.

However my agricultural pilot friends soon told me that going north had its own dangers. Apparently Mauritania to the north of Senegal and Senegal itself both had their share of rebels too and three civilian aircraft had been shot down over the last six months. *Can we go south, Gab?* No, north it is.

My plan was to stay in Dakar for two nights, but there was one thing I had to do which could delay my departure. *60 Minutes* were organising a satellite hook up as part of their continuing coverage and support of my flight and they wanted me to do this from Dakar. They had spent time and money setting it up. I thought at the most that I could be delayed an extra night. Little did I know.

The uncertainty of life is perhaps the only sure thing, or, as Robbie Burns said about the best laid plans of mice and men — of course, that was in the days before women counted — anyway, my best laid plans were blown apart at 2.00 am the next morning when I woke up in a hot sweat

with the most severe stomach cramps imaginable. I was a
victim of the Dakar devastation. I raced to the bathroom and
the smell struck me like a blow.

I think I spent the next two days in a fog of illness.
I was too weak to walk very far and couldn't hold down
anything I ate. I tossed about in my bed with the bathroom
door open and the fog that emanated from it adding to my
total discomfort. I would drift off to sleep only to wake
suddenly from a nightmare in which I was caught in an
Atlantic storm and was pitching around in giant waves. The
Atlantic and my mother became metaphors for each other.
One minute calm weather, the next a terrible raging storm.
I thought more and more of my mother and her life.

She was the happiest I had ever known her when she
was living with Laurie and they were planning to marry and
build their dream home at Clareville. But it didn't work out
that way for her. One day I came home from a weekend
away to be greeted by mother in a distraught state. Laurie
had suffered a massive heart attack and died. He was only
forty-four. From then her life flew apart, and, whatever
stability she had had, evaporated as she started drinking again,
more heavily than ever. With her drinking came the erratic
behaviour that makes life a misery for anyone associated with
the family of an alcoholic. Fortunately my sister Sheri was
at boarding school and, though she came to visit us, was
not subjected to the constant nerve-stretching drama that
became a daily part of my life.

Living with an alcoholic is a terrible experience for
another person. There is no stability or freedom to plan.
There was always a sword hanging over my head and I was
constantly waiting for the next drama to unfold. My mother
would use anything to bend me to her will and often
threatened to take her life to get what she wanted. I lived
in a confused and frightened world.

One time I was asked to go to the movies with a group
of my school friends and my mother gave her permission.
On the Friday night I was getting ready to go when she asked
me what I was doing. I told her I was going to the movies

and she had given her permission earlier in the week. That wasn't good enough for her, however, and she forbade me to go. I begged and pleaded, which did no good, so eventually I rebelled and told her I was going anyway. She pulled out her trump and threatened to kill herself if I went. By this time I was full of her threats and stuck by my decision.

Of course I sat through the movie wavering between feelings of guilt, remorse and rebellion and didn't enjoy it at all. When I got home the house was empty. My stomach dropped as I read the note that said my mother was at her friend's house. She had stuck her head in the gas oven, after first phoning the friend, who had come over and saved her, of course. I was a mixture of emotions, from rage at my mother to guilt at myself. It was a horrid period of my life and one which made my teenage years very stressful.

Many years after, following the failure of my first marriage, when I was a single mother trying to raise Mimi and earn an income to buy a home, my mother came to live in the same block of units as me in Watsons Bay, which is a beautiful part of Sydney.

I had a good job as a medical detailer with a pharmaceutical company and was in the process of making a life for myself and my beautiful daughter. I enjoyed having my mother's company and the help she gave me with Mim, but she soon reverted to her old self and began interfering in my life, and in fact her interference caused trouble with some good friends. I was also frightened of the influence she might have on Mimi. I came to the end of my tether with her and sold my apartment and bought a small cottage at Diamond Bay. I told my mother I didn't want to see her again.

I went ahead with my life and put my energy into buying my house. It was during this time that I began to learn to fly. Part of my job included a weekly visit to the South Coast below Wollongong. I kept driving past the Albion Park airstrip which advertised flying lessons. What the heck I thought one day, and so began the journey that had me lying distraught in a hotel cabin in Dakar.

My mother's influence on my life was very strong and

though I grew up hating many of the things she did, I continued to love her deeply. The hardest thing I ever did was enforce the ban on contact with her. However it was essential for my survival.

After two years of silence I got a phone call from my mother, she had been ill and was diagnosed as having little more than a year to live. She moved in with me and that time was the happiest of my life. We finally established a relationship based on the love we felt for each other and, though her impending death hung like a cloud over us, we refused to let it interfere with our new found association. She insisted I keep up with my flying, and at that time Neville and I were thinking of marriage.

The way my mother faced her death gave me a deep understanding of courage. Every wrong she had done me, every transgression was wiped out by the spirit she showed at that time. I was so grateful for that time we had together. And it was the recollection of my mother's courage that gave me the stimulus to get up out of my bed and get myself moving again. I was still too weak to fly anywhere, but had to wait anyway for the *60 Minutes* satellite hook up. But that meant going to the television station.

Senegal is another country where rule is gained and maintained by armed might and although the army was not in evidence as much as the army in Surinam, when I got to the television station I thought I was entering Fort Knox.

It was like an armed camp with high barbed wire fences and gun emplacements all over the place. Apparently everytime there is a coup attempt the first target is the television station. I suppose everyone thinks if they control the propaganda machinery they control the country, and those who have already been there and done that know the importance of defending their gain, ill gotten though it be. Pity the people who have to live under such constant threat.

The television station smelled worse than the town and I began to feel ill. Fortunately I found out that there was a new television station some distance away which had just been built by the Japanese and it was from there that my satellite link to Australia was to be made. The world is full of inconsistencies and incongruities and one of them certainly was that new television station in Senegal. It features every hi-tech device that Japanese ingenuity has produced and has studios and control booths that are so up to date I doubt their like exists anywhere else. It was more like a space station than a television station. And all of this alongside the squalor, poverty and degradation of the local population. Some equations never add up, no matter how long I look at them.

The *60 Minutes* link went ahead very well. Mimi, James and Neville were in the studio in Sydney and it was like being in the same room with them. This was so different from being on the phone — to see the faces of my children and the expressions of joy and love as we talked. There was no nation-wide audience as far as I was concerned, just me and my kids.

I was surprised that Neville was there. I knew Mimi and James would be there, but when Jennifer Byrne tried to angle the story around to the possibility of Neville and I getting back together I could see they had roped him in, probably hoping that we would make declarations of love for each other on national television. We didn't.

The thrill of seeing my family again was another spur to me getting over my stomach problems and getting on with my journey and home. I eventually decided to try some food and Souheil, a French-speaking Lebanese, took me to a restaurant on a pier over the Atlantic. I could see the water was full of fish, with an occasional shark fin breaking the surface. I definitely didn't want to go down in those waters. Souheil told me of an island not far off the coast where the natives dive for pearls. Almost all the divers have limbs missing. What a way to earn money.

We went to the local market and I found myself literally being pulled by excited stallholders to view their wares. I started to feel ill again and Souheil took me to his home to meet his sister. They had both escaped from Beirut and though members of their family were still there, they knew they could never go back.

The meal and the market combined to bring on another bout of diarrhoea and so I drank bottled water for the next 24 hours. But I knew I was never going to get better staying in Senegal, so I decided to leave the next morning. I had been in Dakar a week.

I got up at 5.00 am and though I'd fasted for 24 hours I still had severe stomach cramps, but I was determined to get out of there. I could have stayed six months and still been feeling ill. I got myself to the airport and filed my flight plan for Casablanca.

Souheil met me at the airport and gave me a beautiful wooden carved salad bowl which his sister bought for me. It was a truly touching gesture, and as I stowed it away I thought it could be a good luck charm.

I said goodbye to Souheil and thanked him for his many kindnesses. I climbed into my little plane and at last got away from Senegal. It was very hot in the cabin and my plane laboured to gain height. I felt it was not performing well. Perhaps it was in tune with me.

11: DAKAR TO IRAKLION

I had my heart set on seeing Casablanca. I suppose it was
the Humphrey Bogart/Lauren Bacall movie that brought out
the romantic in me, however the fates conspired against me
as the cloud base at Casablanca was very low and I found
out at the last minute that they didn't have the right fuel
for my aircraft. I settled on Agadir, a town that was closer
and gave me a shorter flying time, and the weather was okay.
My flight path was north up the West Coast of Africa before
turning inland. I was flying over guerilla country and initially
decided to fly some distance out to sea, but two things
occurred to make me change my mind.

Visibility was very poor because of the Sahara desert
dust, and I had to hand fly the plane, even though I was
still feeling rather weak and dizzy. The other thing that
convinced me to fly closer to land was my sighting of a school
of sharks in the water below me. They were big and the
water was frothing with fish leaping to escape them. I judged
one monster to be six metres long. I felt like I was between
a rock and a hard place. Over land I had the possibility of
being shot down by some trigger-happy guerillas who
couldn't have cared if I was a woman flier from Australia
with no interest at all in their insurgency; over the water
I had the possibility of losing my engine and being dinner

for some voracious Atlantic sharks. I certainly didn't have the stomach for over-water flying.

Let's chance the guerillas, Gaby. My thoughts exactly.

I decided to hedge my bets and flew along the coastline — one foot in the water and the other on dry land. It was sand, sand and more sand, and I didn't see too many signs of habitation, just sand as far as I could see, broken occasionally by a meagre cluster of huts.

Along with my Omega system, the plane's auto pilot had not fully recovered from the battering it took during the electrical storm. But unlike the Omega, it did partially work. I soon found out that it would hold the heading all right, but it wouldn't hold the altitude setting, which meant I was constantly making height adjustments.

I thought about the French Canadian agricultural pilots I had met in Dakar and what a different and interesting life they led. They were part of the United Nations Food and Agriculture Organisation's attempt to control a locust plague, which they told me was of Biblical proportions. Hunter, the sole American in the group, delighted in telling me stories of primitive life in the interior of Africa. Some of his stories were fascinating and some were definitely not to be repeated at a dinner table. Seeing my family on the satellite hook-up and the times I spent with that happy-go-lucky group were the highlights of an otherwise depressing time in Dakar.

I had gained the impression that women were not very well thought of in Moslem countries and it was borne out on my reception at Agadir airport. I went through the formalities of landing and the airport officials were unfriendly and aloof. I reminded myself to take care of how I behaved and I showed an undue amount of deference to these men. It struck me that the cause of women's liberation is a long way from realisation in fundamentalist Moslem countries.

I caught a cab to a hotel near the airport. It was a definite fleabag of a place. The only food on offer was a stewed sheep dish with some vegetables floating around in a centimetre of grease. My stomach was calling out for food, so I took some along with some dried bread and a beer. I judiciously selected pieces out of the dish. I don't think it would have mattered how judicious I was. I lost it all anyway. As I planned to leave early on my next leg to Algiers, I took myself off to bed.

It took me an hour and a half the next morning to go through the country's departure formalities. It always took a long time to go through the procedures, but this was excessive. Usually, I would arrive at the airport for departure, having carefully redone my flight plan at the hotel, and proceed to go through formalities such as customs and immigration. And I would pay any necessary fees. Then I would make my way to the weather office to get a forecast. After the weather I would go to the briefing office to submit my flight plan. When finished with all the formalities, I would load my baggage, although I tried not to take much from the aircraft as it was awful to lug it from place to place because trolleys were not usually available to me, and I sometimes had to wander kilometres before actually getting to the plane.

I was also told that I should wait until permission to enter Algerian airspace came through. I didn't want to wait and asked if it was possible to take off and get permission along the way. They didn't like the idea, but eventually agreed to go along with it. I suppose they thought it was okay to get rid of me as long as I was someone else's problem. The permission came through without any hassles and I climbed to 12,000 feet to cross mountains west of Algiers.

My experience of Algiers was vile. I was accosted by two conmen outside the airport who wanted to change money for me. I was tired and mentally exhausted and on the promise of them taking me to a hotel and picking me up the next morning I gave in to their continual badgering and changed $100. I knew I was being conned but didn't have the strength to reject them.

They took me to the Hotel International. It was revolting and seemed more like a prison than a hotel. It was badly built and badly maintained and very dirty. The people in the hotel were men and it was obvious that they didn't like women. They sat in the cafe and joked with each other and either looked at me as though I didn't exist or with looks that were so obviously lecherous that I wished I wasn't there. They clearly preferred each others' company and many of them held hands.

I asked the guy on reception about making phone calls to Sydney and he made a big show of taking down the numbers and told me he would put through the calls to my room, which I thought was a very civilised service. I managed to stay up until 9.00 pm but my exhaustion finally got the better of me and I went to sleep without receiving any of the calls.

My next leg was to Tunis in Tunisia and was only a short hop, so I decided on a leisurely departure. This was just as well as this time it took me three hours to get through the airport departure routine. I was fuming. Before taking off I noticed an oil leak on one blade of the propeller and an engineer from Air France looked at it for me and told me that it was no problem, so I took off.

The country was a lot more civilised and populated and the Tunisian control kept badgering me to report my position. I was not used to this frequency of reporting, which is a good idea, but at that time it really annoyed me. I was used to flying long stretches without any radio communication and to be suddenly confronted with constant demands to report my position seemed like an imposition.

There was a build up of stratocumulus over the mountains west of Tunis and I was bounced around. As I came through I requested a descent to 9000 feet, but they kept me hanging up there at 12,000 feet for an inordinately long time. Flying over 10,000 without oxygen isn't a good idea, particularly if you're nor feeling well.

About 9 nautical miles out the airport was clearly visible and I requested permission to make a visual approach. I was directed to make a very lengthy instrument approach, which seemed very silly to me. All the airports since reaching South America had made me do instrument landings even in visual conditions. At least I got a lot of practice.

Tunis was a breath of fresh air. The airport staff were friendly and efficient and though I couldn't get Avgas there, they arranged to truck some drums in from an aero club at a nearby airport. Getting fuel was becoming a problem. Because I was an international traveller I had to land at international airports and in many countries the international airport is only set up to service jet planes. Anyone landing in a piston-engined airplane was an anomaly. It would have been easier if I could have landed at a local strip, but of course these were not official entry ports.

My plan was to stay overnight and the next day head across the Mediterranean to Crete, however the airport meteorological office gave me a gale report over Crete, and this was confirmed by Boston Bob when I phoned him from my hotel. I ended up staying three nights and I was thankful that this happened in Tunis and not Algiers.

Tunis is a cosmopolitan place, having very much the air of a French city, with its wide, tree-lined boulevards and tree guards and *pissoires* that reminded me of Paris. It also has the added attraction of a history that stretches back to Roman and Carthaginian times. The hotel was of an international standard with clean rooms and a restaurant that knew how to serve a decent meal. I was still suffering the lingering effects of my stomach disorder but the food didn't upset me, which was a good indicator of its quality and my increasing recovery.

I used a lot of my time in Tunis to rest and regain my strength. I really shouldn't have left Dakar feeling as I did, but I had just wanted to get out of Senegal. I also caught up on my phone calls, as my family were beginning to get concerned that they had had no contact with me since leaving Agadir, Morocco.

One of the airport officers offered to show me the markets in the old city, so on my first morning, after making sure the fuel had arrived for my plane, I set out with him on a journey that took me back in time to a way of commerce that is making a resurgence through the markets in our cities and towns.

I don't think I have ever seen a market as smelly,
claustrophobic, noisy or exciting as the market in Tunis. It
is built in narrow underground tunnels that have sky-lights
to let in the natural light. I was still unwell and my fascination
was interspersed with attacks of nausea as I succumbed to
the excessive heat. The body heat of so many people, added
to the heat and humidity of the day, produced a miasma of
body odour which mingled with the market smells and did
not have a good effect on me in my sensitive condition. I
suffered waves of giddiness which forced me to stop and
pause before going on with the tour of the attractions.
Metalsmiths and cobblers pounded away with their hammers
in a cacophony which added to the volume of noise from
the people and produced a confusion of sounds which to
my ear sounded like bedlam. I bartered for a lovely ceramic
plate and apart from the bouts of nausea really enjoyed this
visit.

On my second day, and still forced to stay by the gales
sweeping from the south across Crete, I negotiated with
Abdul, a cab driver outside the hotel, for a conducted tour
of some Roman ruins a short distance out of the city,
overlooking the Mediterranean Sea. Abdul was a character
and a fount of information. He was also a pleasant rogue
who did his best to part me from as much money as possible.

There were two old men at the ruins who Abdul told
me were the official guides and for a small baksheesh would
give me the *tour de luxe*. Those two guides were more
ancient than old and yelled at me through their toothless
mouths until we agreed on a fee for their services. When
this was achieved they became extremely chivalrous and
showed me through a colliseum and the remnants of some
Roman baths. I also drank from a well that is over 2000 years
old and imagined the women of those ancient times coming
to this well to fill their ewers and go back to their dwellings
to carry out the household drudgery that seems to be the
lot of women in most of the countries in this part of the
world.

Before leaving the site, the two ancient guides offered

to sell me some genuine Roman relics that 'had recently been unearthed'. By now I was getting used to this way of doing business and, knowing the 'relics' were made the week before, decided to haggle with them to see what sort of price I could get them down to. I enjoyed the exchange as we bantered back and forth for these 'genuine' artifacts — some coins and a lamp. I'm sure they still got the best of me, but I gave a good account of myself.

On the way back to the hotel Abdul took me to the oldest mosque in Tunisia. It was very old and very beautiful and I was able to look inside and see the exquisite tiles and lattices.

The weather report that night was for lighter winds over Crete so I planned for an early take off for one of my last major water crossings. One thing I had learned in the Moslem countries was that I had no need to book early morning wake up calls. The muzzeins calling the faithful to prayer at 4.00 am every morning made sure I awoke. It was a beautiful experience lying in my bed, listening to the rising and falling of the muzzeins' voices and I often had to pinch myself to make sure I was not part of an Arabian tale.

Tunis was the most outstanding part of my stay in North Africa and is a place I would not mind going back to. The airport formalities were efficient and the weather information excellent. I could usually understand the weather forecast, wherever I was, even if language difficulties existed, because the format for giving the information is universal. However, sometimes I was frustrated because it was my habit to ask in a simple fashion, 'How is the weather for my route?' before reading it. In normal circumstances, a friendly, helpful meteorological officer, who speaks the same language, of course, would just say, 'bloody awful', or 'not bad', or 'great'. These words said volumes to me. Here in Tunis they were very helpful, too. I got away at my planned time of 8.00 am, which was another North African milestone.

My flight path was to take me over Malta and across to Crete. The weather was good, as forecast, and, as I looked down on the calm Mediterranean below me I mused about

PREVISION DE VOL

Vol n° PVT (private)

Trajet $LGIR$ (IRAKLION / CRETE)

Communiqué par le Centre météorologique de : $DTTA$

à $Atoo$ T.C.U.   le 07 .-10-₁₀ 89

```
ZCZC 280 10000
FCGR31 LGAT 070000 RRMYVHYKTCAVOKGRADU060836016KT99992CU02048C
030=
LGTS  0110 VRB08KT  CAVOK=
LGAD  0110 VRB08KT  CAVOK GRADU 0608 34012KT=
LGRP  0110 33016KT  9999 2CU025=
LGIR  0110 34016KT  9999 2CU020 4SC030=
LGKR  0110 VRB08KT  CAVOK GRADU 0608 34012KT=
LGKO  0110 36018KT  9999 2CU025 4SC030=
LGSA  0110 36006KT  9999 3SC030 GRADU 0608 36015KT=
LGLM  0110 05008KT  9999 2CU025 GRADU 0608 05016KT=
LGSM  0110 36015/25KT 9999 2CU025=
```

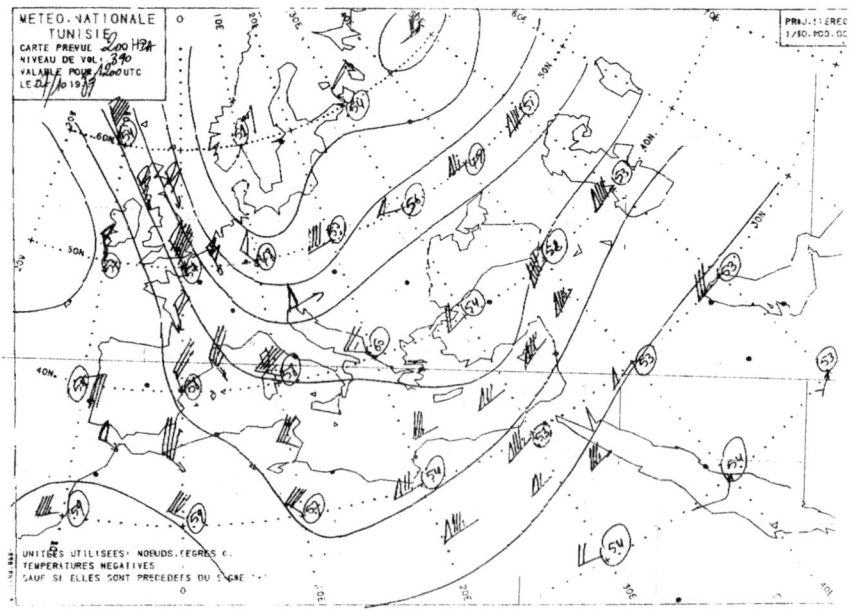

*Part of Meteorological Report from Tunisia to Iraklion.
Meteorological reports are standard around the world,
and are in standard codes*

the moods of the various stretches of water I had crossed
and how the people and lands that skirted the shores seemed
to reflect the characteristics of the sea or ocean.

The Mediterranean seemed calm and benign, though
likely to produce an unexpected storm. In contrast, the South
Atlantic was definitely tempestuous, given to violent rages.
It was big and heavy and brooding. The Caribbean was
amiable and inviting, but held a veiled danger which might
erupt in the hurricanes that swept in from the Atlantic. It
also had an air of mystery. The Pacific was vast and generally
kind, but always held the threat that it could swallow up
people, land and airplanes.

I flew across Malta and a controller there, in a kind
gesture, told me that an American fleet was patrolling in the
waters I was going across, which was an indirect way of
telling me that if I went down rescue wouldn't be far away.
This was reassuring because I had requested, through Malta
control, permission from the Greek controllers to make a
more direct flight path, rather than the triangular path the
instrument charts laid down. Their answer came back that
it was okay, as long as I realised that my safety was my
responsibility.

This was good news to me, as it meant I could save
time and fuel, and although my crossing of the Mediterranean
was only six hours, and relatively minor compared to the
major crossings of the Atlantic and Pacific, it was still a long
crossing and anything that cut down the time over water
was a blessing.

It was during this crossing that I again heard the voice
of Julian from Portishead over my radio. I quickly made
contact with him and gave him an update of my journey and
thanked him profusely for his help on that dreadful night
over the Atlantic. I felt myself beginning to cry and ended
the transmission just before a deluge of tears overtook me.
All the frustration and fear and loneliness that I'd experienced
during the night of storms came back to me and I just let
go and indulged myself in the best crying jag I'd experienced
for many years. It was good to get it out and I felt much

lighter afterwards and even laughed at the idea of letting go and allowing the plane to look after itself.

Without the Omega I was reliant on picking up the NDB on the western edge of Crete, and it came in right on cue. Things were going well for me. I climbed to 12,000 feet to keep well above the mountains as I crossed from west to east of the beautifully spectacular island. The mountains edged right alongside the sea and, though there was cloud around the mountains, I had a wonderful view.

As I approached Iraklion on the eastern edge the controllers kept me at 12,000 feet and had me descend over the sea to the east and make a complicated instrument approach which brought me in on the east-west runway, looking right into the lowering afternoon sun. I was temporarily blinded, but managed to stay on course and make a reasonably proficient landing.

The airport was a joint civil-military one and was ringed with gun emplacements, which were manned. The Greeks seemed fearful of an invasion from Turkey and made sure the airport was well defended in case any such emergency arose. I needed to organise fuel and an oil change, so one of the men from customs arranged for me to fuel at the aero club, and for the club's engineer to do the necessary oil change. As soon as the engineer saw the *60 Minutes'* cameras mounted on my plane he told me to quickly dismantle them, wrap them up, then hide them away, and he helped me to do it. We had just carried this out when an army officer and some soldiers came over and inspected my plane.

In all the countries that I had so far visited, even those with a strong military presence, none had shown the slightest interest in the *60 Minutes'* cameras. Most of the time I was so engrossed in my flying that I forgot to turn them on, more so after the Atlantic storm when I had to pay extra attention to my navigation and flying because of the damage to the auto pilot. It seemed strange to me that the much heralded cradle of democracy would find it necessary to show so much military muscle.

The engineer very quickly attended to my oil change

and invited me to his home for dinner. I asked him if he could take me to a reasonable hotel.

The hotel he took me to was in the older part of Iraklion and called the hotel Atlantis. I checked in and, as soon as I got to my room, I phoned Bob Rice, my dependable weather man in Boston. Bob's news was not very good. Get out early the next morning for Cairo. Bob's prediction was that the weather I was experiencing was only a temporary lull and the gales that kept me in Tunis would return the next afternoon. I groaned inwardly and rebelled. I was looking forward to a day off from flying as I was still recovering from the stomach bug and the flight from Tunis had worn me out.

Better do what Bob says, Gaby.

Shut up. But I would have served myself better if I had heeded the advice from Bob.

Johnny, the engineer, came by at 8.00 pm and drove me to his home, where Helena his wife had prepared a beautiful Greek meal. We were joined by his two children and I enjoyed a very convivial evening with this family who so very generously shared their home with me. The meal was special. I have always been fond of Greek food, but it had never tasted like that meal. We had taramasalata, tzatziki, homous — the whole bit. My stomach relished the food and I knew I was recovering from the Dakar horror.

Well, the rebel in me won out and that night I went to bed in my hotel determined to sleep in and take off at my leisure. I surfaced mid-morning the next day and knew as I drew the curtains aside that I wouldn't be leaving Iraklion that day. The wind was blowing a steady 40-50 knots with stronger gusts. It came from the south-east and had the smell of desert in it. Visibility was very poor.

'All right, girl, you might as well enjoy yourself.'

The Hotel Atlantis was not very large, although it was of international standard. I was lucky to find a room, as the hotel was full of German tourists who, along with Scandinavian and English tourists, make a mecca out of Crete. Their influx onto the island has strained a lot of resources and the airport is reduced to chaos if three planes arrive at the same time. Apparently the wait for processing almost drives most tourists home. I've heard many criticisms of the inadequacies of Sydney's international airport, but it's a breeze in comparison to Iraklion, although a new airport is under construction nearby.

I thought I would go and see some of the sights and, after a late breakfast, walked outside and approached a cab driver leaning against his cab. It was obvious that he was fed up with tourists and looked at me with some disdain. When I started speaking English in my peculiarly Australian accent he seemed to warm to me a little and agreed to take me on a day's sightseeing and included a meal at his house as part of the 'package' deal.

He opened the back door of the cab for me and as I was thinking what a gentleman he was, I felt a very firm pinch on my bum. My hackles started to rise and, as I turned to face him, the grin on his face showed me that he was expecting such a reaction, so I quickly changed tack and, in a very quiet voice, told him not to be silly. I surprised myself at my calmness and I definitely surprised him. I was careful around him from then on, but he backed off and knew that I wasn't going to give him a chance. I must admit that it's the first time I've been pinched on the bum and I don't think I could grow used to the custom. I much prefer a gentle pat.

He took me around the town and down to the old harbour, which is a safe anchorage in all kinds of weather, yet on this day there were waves chopping up its surface and it reminded me that I should have taken Bob's advice. It looked like the wind was going to blow for some time and I would be in Iraklion for more than one extra day.

My cab driver was a very interesting character and, as promised, took me to his home for lunch. His English wasn't very good and I knew no Greek whatsoever, so we conversed through signs and gestures and broken English. His name was unpronounceable to me so I christened him Charlie, which he appeared to accept with equanimity. The garden in his home was his pride and joy and he insisted on showing me the wonderful array of salad vegetables he was growing there. He picked some, collected some bread, cheese and wine from inside the house and we sat down on his verandah and had a really beautiful, fresh lunch. It was topped off with thick Greek coffee.

He invited me inside, but I made sure I stayed outside, so he cleared up, then drove me out into the countryside for two or three hours. We stopped at a small village along the way and I must admit I really enjoyed it.

That night I went to the hotel restaurant for an evening meal. It was full of German tourists, who seemed to have the staff running around in all directions. There was no *major domo* so I stood in the entrance for about ten minutes waiting for someone to notice me. At last an old Greek man came up to me and, taking me by the hand, led me to a table where five German people were already seated. There was a spare chair so he sat me down and then went off to find a waiter for me. I nodded my head in greeting to the other people at the table who gave me a cursory glance and then ignored me to pick up their conversation in their guttural language. They continued to ignore me throughout the whole meal. I overcame my feelings of rejection and ended up thinking how poor in spirit those people were. An old saying of my mother's came back to me: 'Manners maketh the person', and while I know that's not entirely true, poor manners generally do reflect poor people. I retreated to my room as soon as I finished my meal.

My decision to rebel against Bob's advice cost me five days in Iraklion. Days spent waiting for the wind to blow itself out. Days that were generally frustrating and boring. I twice tried to leave, but the people at the airport talked me out of it. They told me that the turbulence around the mountains was dangerous, particularly at the angle the wind was blowing. The chop in the protected harbour had turned into metre high waves. It seemed I had set myself up to relearn the value of patience, which I had forgotten. Once I accepted that I would have to sit it out, the weather started to improve.

I was in touch with Bob every day and at last the word came through. Tomorrow's wind was expected to drop to 25-30 knots and, more significantly, it was easing towards Cairo. I planned to leave the next day.

Bob's prediction was accurate and the next morning I woke up to a more manageable 25 knot wind. I packed and walked down to the restaurant for breakfast and, to my delight, it was empty. The German tourists had gone. As I walked through the room I spotted one other couple, so I chose a seat several tables away from them, and, as I was checking out the menu, I was distracted by the conversation the couple were having. It wasn't really the conversation, as I couldn't hear all that well, but there was something familiar in the way they were talking. Then all of a sudden it dawned on me. They were Aussies! The last Australian voice I'd heard was on the phone in Belem, when I was talking to the Australian ambassador in Brasilia. I gathered my courage and walked to their table.

It was almost like a homecoming. My new friends were a middle-aged couple from Victoria. They were on the final leg of a world tour and anxious to get home. They invited me to join them at their table and we talked for some time, but it soon came time for me to go. I really envied them as I rode in the cab to the airport. Home for them was only a day or two away, with a Jumbo to whisk them there. As for me, I had a long way to go. Home seemed like a dream. Oh, how much I wanted to be there.

As I took off from Iraklion airport and headed across the south-east tip of Crete, climbing into the strong headwind for altitude to clear those mountains standing magnificent in the morning sun, my spirit picked up. At least I was heading in the right direction.

Right on, Gab.

Yeah. Right on!

12: IRAKLION TO BOMBAY

Home. That magic word that meant so much to me. Some words from a poem by the American Robert Frost kept running around my head as I flew across the Mediterranean to Egypt:

'But I have promises to keep,
And miles to go before I sleep,
And miles to go before I sleep.'

Of course, those words are from a poem entitled *Stopping By Woods On A Snowy Evening* and I was a long way from any woods, and a long way from any snow, but the sentiment was the same — I was on my way home. So much of my journey was completed. All the major water crossings were done. The only significant water to come was the Bay of Bengal, and after that it was a downhill run, island hopping to Australia.

Visibility was poor. There was a lot of dust in the air, making it difficult to see for any great distance. I crossed the coast of Egypt near Alexandria and flew across the verdant countryside of the Nile delta region. It was so rich-looking and in direct contrast to the approaching desert. Cairo was still a way off, but the Egyptian air traffic controllers were determined to let me know they had me under their control. At least their incessant chatter made it very lively.

Visibility became progressively worse the closer I came to Cairo. I suppose the desert dust mingled with the pollution of the city to produce an effect that was like flying through something solid.

Cairo control seemed to be in a panic situation, which I soon learned was fairly normal behaviour. There was a lot of traffic in the air and they put me in a holding pattern 25 miles north of the city, which I had to maintain for 20 minutes. I was beginning to despair of ever getting to land, when they gave me the okay to proceed to the international airport. I was on instruments, which was just as well. The haze became progressively thicker.

The closer I got, the more excited and confused the controllers became. They kept me circling for a while and then gave me the clearance to land. As I came in I noticed a 747 waiting for me to land so it could take off. I was 100 feet above the runway when a controller's voice that seemed on the edge of panic came over the radio. 'Victor Hotel Golf Kilo Foxtrot you're at the wrong airport. Go around. Go around. Go around.' Oh, God, what's going on, I thought as I retracted my wheels and flaps and began to climb.

Just then another voice came on. 'Victor Hotel Golf Kilo Foxtrot you are clear to land.' Oh Shit! What is this — the bloody keystone cops!!!!!

I was fast running out of runway so I got the wheels and flaps down and made what I thought was a hasty landing, but then the 747 captain came on my radio and in a very American voice said, 'Lady, I don't know how you did it, but that was a beautiful landing.' I muttered my thanks and taxied off the runway to let the big plane take off. I don't know if that anonymous captain had any idea of the soothing effect his compliment had on me. I was infected by the panic that the controllers operated on and I was in the process of beating myself up for fouling up the landing when those words of praise crackled in my headset. It was like manna from heaven.

Cairo international is an enormous airport and I was directed to a parking bay almost 2 kilometres away from the runway I'd landed on. I came to a stop, shut the engine down and opened the door. Two Egyptian men in uniform approached me and introduced themselves as handling agents, and offered to act for me. It is usual practice for pilots in strange countries to utilise the services of handling agents, who make it easy to get fuel, weather reports, lodge flight plans and generally find your way around a strange place.

Excited voices seemed to be the norm when Egyptians converse — just like the air traffic controllers, these men obviously didn't know the meaning of a *sotto voce* and conducted their spiel in a volume that made me feel like they were yelling at me. Their price was $200.00, which I agreed to, even though it did seem somewhat excessive.

Okay, I said, first off I need Avgas. I want to fuel up so I can get away early in the morning. They carried two-way radios and one of them yelled instructions into one of these and it wasn't long before a fuel truck pulled up alongside my plane. The guy in the truck jumped out and a rapid, high-volume conversation was carried out in Arabic. The fuel man unreeled his hose and I noticed in English below the Arab characters on the tank, the words Jet Fuel. 'Stop! Stop!' I screamed as the guy prepared to fill my plane. 'That's jet fuel,' I yelled at the handling agents. 'Stop him. My plane takes Avgas.' The fuel man blithely carried on and unlocked one of my wing tanks. I ran over to him and virtually had to wrench the hose out of his hands to get him to stop.

At last the message got through and the potential disaster was aborted. The two-way radios came out again and the handling agents carried on another screeching conversation with someone on the other end. They finally brought their attention back to me to tell me that their boss also owned a light plane, which he kept at another airfield in Cairo. They suggested that I fly my plane there. I refused, because I didn't know my way around and I doubted if the controllers would let me do this, as I didn't have a flight plan. Besides, I needed to go through customs and immigration and Cairo international was the only entry port.

I told them they would have to ship some fuel over in a truck for me. I just needed enough to get me to Luxor, which was on my flight path — 30 gallons would be enough. I told them I would stop there and get more fuel.

They told me that it could be done the next day.

'Oh no. You get it organised today,' I said. 'I need to leave here early in the morning, as my clearance to fly over Saudi airspace cuts out tomorrow.'

I was warned by QANTAS staff in Sydney that it was imperative that I not foul up my clearance to fly across Saudi Arabia. I was lucky to get one and if it expired they doubted I would be given another.

There was more yelling on the two-way radio and the handling agents said it could be done. It might cost a little more money. That's okay, just get it done, I told them. And that turned out to be a very big mistake.

All this was settled beside my plane and we were a long way off from the customs and immigration inspection point. Out with the two-ways, and the next minute an airport bus pulled up beside us and we climbed in. The bus ride was a hoot. There were four Egyptians and me and I expected at one stage for the bus to stop and the four guys to get out and fight on the apron. They were yelling and screaming at each other and yelling into two-way radios. I started to get uptight and then all of a sudden I saw the funny side of it. I was watching the Egyptian version of the Three Stooges. I sat in the back of the bus and started to giggle.

My handling agents got me through customs and I was outside the airport in half an hour. They drove me to the nearby Moovenpick Hotel, which is part of a European chain. It was a beautiful hotel of truly international standard. The handling agents agreed to pick me up the next morning at 7.00 am. I wanted to be sure of getting to Bahrain before nightfall the next day.

As I entered the hotel a wedding was in progress in a reception room off the foyer. It was a very colourful affair and I wish I could have stayed and been part of it, but I was feeling tired and wanted to just have a snack in my room, phone home and get to sleep.

The next morning I was on the hotel steps at 7.00 am. Needless to say my friends didn't show up until 8.00. By then I was fuming. I tore strips off them as I climbed into the car. Their negligent attitude was costing me valuable time. I wanted to call into the meteorological office to get a weather report, but they insisted on mucking about and completely ignored my request. We got to my plane at 9.00.

'We've put the fuel in,' they proudly told me. I didn't like this. Fuel is a critical element to a pilot and I always liked to be there to check when my plane was fuelled and to draw samples from the tanks to check on water or other contaminants in the fuel. They also told me they had submitted my flight plan and checked on the weather, which was fine.

I just didn't trust them and, as there was a pilot from a private jet nearby, I walked over to him and asked him if he had a meteorological report for the Luxor area. He told me he did and that the report was for fine weather. I then walked back to my plane and my handling agents handed me a bill for $1000.00.

I exploded. They yelled and started gesticulating and making veiled threats. I accused them of robbery and just about every other crime under the sun. They reckoned they could justify the costs and proceeded to go through a lengthy song and dance about how much each item cost. They pulled out a sheaf of documents written in Arabic to back up their claim.

They had me by a place I wished I could have had hold of them. I knew my choices were limited. I could stay and fight it out with them and kiss any chance of flying across Saudi Arabia goodbye, or I could shut up, cop it sweet and get on with my journey.

Know your priorities, Gaby.

I paid up and left Cairo with a bad taste in my mouth. Many people had told me about the friendliness of the Egyptian people. So far I had only experienced the parasites.

Up in the air I was free of the atmosphere of Cairo. I washed the rotten taste out of my mouth and settled down to flying the 270 nautical miles to Luxor.

It was truly arid beneth me, kilometres on kilometres of sand and not an oasis or a camel to break the scene — endless sand that baked at 50°C. My strong imagination came into play again and I wondered how I would survive if I had to land in the nothingness below me. I know I'm prone to melodrama at times, but looking down on the featureless sand produced a tension which, added to the hassles in the airport, produced knots in the pit of my stomach. How was I to know what type of fuel, and how much, they had put in my tanks? I did some deep breathing and prayed to the desert gods to look after me.

I crossed the Nile ten minutes out of Luxor. The contrast was dramatic. One minute sand, the next the lush growth of the river basin that produced the mighty civilisation that once dominated the known world.

Landing at Luxor was difficult. It was intensely hot and visibility so poor that I couldn't see the runway. It was by now after 11.00 am and my goal of crossing Saudi Arabian airspace and safely reaching Bahrain seemed to be slipping from my grasp. 'Thanks to those dick heads back at Cairo', I muttered to myself, as a small fat Egyptian man came up to my plane and introduced himself as the Luxor representative of the firm of handling agents that had a loan of me in Cairo. It seemed the $1000.00 I paid over included the services of this man.

'Okay', I yelled at him, 'get my plane fuelled up'. He was behind the eight ball with me from the start.

So, while he fussed around organising the fuel with the inevitable walkie talkie, I spoke to a local pilot, who told me that no one flies in that country in the latter part of the day. The sun is so fierce and the ground temperatures so hot that the air at sea level becomes less dense and engine performance suffers. It is like flying at 10,000 feet and with a full load I would need a lot of runway to get off the ground.

The handling agent came over and kept urging me to

go. I quickly had him figured out. He just wanted to get rid of me as I might cause more work for him. He couldn't have cared less if I made it or not. I hurled a few well chosen Aussie colloquialisms at him and walked over to the control tower to see what could be done about my permit to cross Saudi Arabia.

I think I walked into the real Egypt and met some of the real Egyptians when I walked through the doors of the old tower at Luxor airport. I was immediately made welcome and after listening to my predicament, the senior controller invited me into the inner sanctum to have a cup of tea while he attempted to solve my problem. The building and its equipment were very old and the three guys working there were bright, cheerful and happy. They obligingly swept some rubbish aside and put me on a chair, and then proceeded to brew up with an old kettle bubbling away on an ancient primus stove. The tea they gave me was black, strong and sweet and came in a rather dubious-looking chipped enamel mug. I didn't mind, I knew I was amongst friends.

I think the cup of tea I had in the tower at Luxor was the best cup of tea in the whole flight and every bit as enjoyable as the wines I'd sipped in other parts of the world.

I told the guys there about the hassles I'd had with the handling agents in Cairo and the comments they made about the local representative confirmed my suspicions that he was only interested in getting rid of me and keeping my money. They also considered that handling agents duped their clients. One of the guys told me in his accented English, and with a grin on his face: 'They couldn't handle camel shit.' I had to agree with him, and we all started laughing.

The tower chief got through to Jeddah Flight Information Service and through the grace of Allah managed to get a one day extension on my permit to fly over their air space. This was great news and indeed a miracle. The QANTAS emphasis that I not miss this deadline had been reiterated by the Australian Department of Foreign Affairs, whose representative in Riyadh had done a lot of work on my behalf to help gain the clearance. Women are kept in purdah in

Saudi Arabia and aren't even allowed to drive cars, let alone aeroplanes. Foreign Affairs told me I'd been lucky that I'd been given a clearance in the first place and they had insisted that I not miss the deadline.

I walked out of the tower building a new person. I chased up the handling agent and ordered him to take me into a hotel in town. I was determined to get some value for the money I'd paid.

The Hotel Etap was beautiful. A magnificent old colonial building on the banks of the Nile. Gleaming brasswork, which obviously received a daily polish, glinted in the sunlight. My room overlooked the Nile and the Valley of the Kings. It was magnificent and made me want to come back to this sacred city with the ruins of its ancient temples. This is Egypt, I thought, not the turmoil I'd encountered in Cairo.

Surprisingly the handling agent showed up on time the next morning. I wondered if he'd heard from his Cairo colleagues of the bawling out I gave them when they showed up late.

Getting out of Luxor was a hassle and the authorities were unpleasant and rude. Ah, this was more like the Egypt I was used to. The handling agent, who was supposed to smooth the path for me through the mire of officialdom, kept disappearing to go and handle jobs for other clients, which made it necessary for me to go and find him and drag him back. Eventually, I laid down the law and told him to stick with me and get me through. He meekly obeyed.

I took off with a sigh of relief, very pleased to leave Egypt behind me.

I was still well to the north of Amelia Earhart's flight path. She had flown directly across Africa from Dakar and from Assab flew south across the Arabian Sea to the Indian sub-continent. My flight path was from Luxor across the Red Sea and over the Saudi deserts to Bahrain, a small island off the east coast, and from there across a small section of the Arabian Sea to Karachi in Pakistan.

The Red Sea was beautiful to look at. I was told that it was once one of the richest seas in the world, with a proliferation and diversity of marine life which provided the bordering populations with a plethora of rich foods. Unfortunately the spills that have resulted from the huge quantities of oil carried through the Sea have devastated the once abundant harvest.

The western coastal area bordering the Red Sea is some of the most spectacular country I have ever flown over. Although it was arid and obviously desert, tall mountains of bare rock rose majestically from the sand to heights of 250 metres, their surfaces constantly scoured by the blowing sands.

A 20 knot wind was blowing across my path and I was thrown about by mechanical turbulence as my attention was constantly drawn to the grandeur of the vistas below me. Then the country gradually flattened out and became more like the desert I was used to. Groups of nomads with their flocks were visible on the desert floor.

Jack Hobbs, the man who was my navigation teacher at Sydney Technical College during my student days, had helped me prepare the topographic charts for this flight. I was equipped with two types of charts: the topographic ones, which showed geographic details and features of the land I was flying over, and instrument charts, which detailed the navigational aids and tracks, lowest safe altitudes, etc. Contending with two sets of charts in the cramped cockpit and all the other pieces of paraphernalia was an exercise in delicate manoeuvring.

Jack had written in bold pencil on the charts for this area: NOTE! UNDER NO CIRCUMSTANCES MAKE A LANDING IN SAUDI ARABIA!!!!

Jack was one of the many people who cautioned me against any contact with the Saudis. It was the woman thing again. The Saudi's strict observance of the fundamental Moslem laws meant that I would be in trouble if I had to call on them for help. Personally, I didn't think they would treat me in the way my friends thought, but fortunately I didn't have cause to test out my speculation.

One of the most interesting things that happened to me

as I was flying across Saudi Arabia was being treated as if I were male. The controllers in Jeddah who made brief contact with me kept calling me 'sir'. I don't know if this was because they couldn't accept the fact that a woman would (or could) fly a plane around the world, or they genuinely didn't recognise my voice as that of a woman's, or perhaps they decided I must have had balls to fly around the world, and so declared me an honorary man. I chose to believe the latter and chuckled to myself each time Jeddah control referred to me as 'sir'. It was a funny experience, but the reason didn't really matter as it helped get me the essential clearances.

As I drew closer to Bahrain and came under their control, an American controller spoke to me. It was so great to hear myself referred to as ma'am, and we carried on a great conversation.

Visibility rapidly deteriorated as I crossed the coast on a heading for Bahrain. It was very hazy and the nearer I drew, the heavier the haze became. I was flying in real pea soup conditions as Bahrain radar vectored me to my final approach.

Going through the formalities was simple and I was treated with courtesy and respect. The officials were very efficient.

I was expecting to be met by Yvonne Trueman, a 99, but she was called away to the United Kingdom to receive a very special award and asked Brenda Pyle, a friend, to meet me in her place. Brenda and her husband Richard were at the airport and invited me to share their apartment. Richard was a roving correspondent for AAP and Brenda worked as a freelance journalist and writer. Maseeh Shehab from QANTAS was also at the airport. Bahrain is a major stopover for many QANTAS flights and they have a service facility at the airport. The contact with QANTAS reminded me that home was drawing closer.

I stayed in Bahrain two nights and was fascinated by the sheiks and their flowing white robes and headgear. They seem to like women with the blonde, Nordic look and many of them were at the best restaurants with these girls, while their own women were back home in the harem. Apparently they shower these women with jewels, but when the affair is over, the girls are expected to return them. As Marcus Clark said: 'Such is life.'

I was apprehensive as I took off from Bahrain on my way to Karachi, as part of my route was through Iranian airspace. Visibility over the gulf was extremely poor, but any danger passed without incident until I entered Pakistani airspace.

I was approaching Karachi from the West, admiring the beauty of this rugged coastal region. I was on my approved flight plan when the Karachi controllers told me to head back out to sea on a south-easterly course, until I intercepted the regular inbound track which was normally used by jet planes.

It wasn't hard to tell I was back in a Moslem country, as the tone of voice used by the controller reinforced my feelings that women were held in contempt. I objected to this and told them I was a small, single-engined plane, but my entreaties fell on deaf ears. I cursed, because it meant I would have to calculate revised times while trying to locate a position 60 miles out to sea and without the aid of the Omega. It would be just a guess.

I made it and reported to Karachi, who came back to me and told me I was now cleared for the original coastal flight path. Not only had they cost me time, but they had also created a lot of stress for me which added to the tensions of flying over strange waters. I exploded and used several four letter words under my breath and told them I was staying on the sea route. I'd never spoken like that over the radio before, but their instructions were stupid and I was at my wit's end.

I continued on my flight path and, although I could see the airport, I was directed to follow the usual involved instrument approach that took me inland for a long way and brought me onto my final approach directly into the strong western sun. But the landing was executed without trouble.

I was barely out of the plane when I was approached

by two men representing a handling agent. Shades of Cairo, I thought. Their suggested fee was exactly the same, $200.00. No thanks, I told them, I only need Avgas. Once bitten, twice shy, I thought to myself. They were obviously upset by my refusal and stamped off with a real display of bad temper. And they set their revenge for me.

Any angry thoughts I had about the bureaucracy in Brazil disappeared out the window as I battled with the Pakistanis. They made my hassles in Brazil seem like a breeze. After organising fuel, I followed a soldier to the customs office. I'm 1.6 metres and there I was loaded down with my bags, following this 1.8 metre tall soldier who kept marching ahead of me at a military pace that forced me to almost trot to keep up. Straps slipped off my shoulder and I was forced to stop to readjust my load. I berated him about his ungentlemanly habits, but my invective went over his head. 'Hurry up', he barked.

It took three hours to go through the processing needed before I could stand outside the airport. Three hours of frustration. Unfortunately a game of cricket was being broadcast over the radio in the customs shed. I think the game was between Australia and Pakistan. I was ignored as the petty officials listened to the game. I think the handling agents whose services I had refused had been there before me. I'm sure I would have found the situation comic in other circumstances, but again being treated as an inferior species added to my frustration and I crossed words with the officials, which of course made them go slower.

The hotel I stayed in was touted as being of first-class international standard. If that was first, I'm glad I didn't stay in third. It wasn't clean, the cold water tap didn't work, I couldn't make phone calls out and the mattress was lumpy. The restaurant was full of Pakistani men who looked at me with a disdain that was repugnant. Fortunately some Frenchmen saw my plight and invited me to join them at their table. As I sat with them and chatted in French I felt the eyes of the Pakistani men. It was very uncomfortable. Also my companions advised me not to eat any salad vegetables and after Dakar I was wary about the food I put in my mouth. The meal was not very appetising.

I slept fitfully and got to the airport early the next morning to lodge my flight plan and get away on my four hour flight to Bombay in India. It took five hours to get out to my plane. I walked from one end of that airport to another. I certainly got the run around from one petty tyrant, who sent me onto the next, who gave me the run around and sent me onto the next, etc, etc *ad nauseam.*

There were many examples of their ineptitude. Most of the forms I was made to fill out had to be done in triplicate and there was no carbon paper, which meant I had to fill each form three times. I paid one of the airport fees in one place and had to go somewhere else to get a receipt and then go somewhere else to pay another fee.

I did my best to make sure the foul up I'd experienced to my flight plan coming into Karachi wouldn't be repeated. I filed a plan which would take me down the coast to Bombay and asked the flight service people to check that it was okay with the Indian authorities. I was told that it had been approved.

At last I made it to the gate to get access to the parking bay. I was refused entry because I didn't have a boarding pass. I blew up and gave such an awesome display of temper that I was rushed through. Maybe I should have tried that approach earlier.

I dragged myself to my plane in the sweltering 45 degree heat. What a disorganised country. I had spent a total of 20 hours in Pakistan. Eight of them trying to get in and out of Karachi airport.

I got into the air and had just intercepted my outbound track when the Pakistani controllers told me that I couldn't proceed on the track I'd originally submitted and would have to totally revise my flight plan to take an inland route. By this time I'd run out of curses. I almost cried as I got out my maps and slide rule and tried to fly the plane while marking out my revised route.

It's okay, Gab, you can handle it. Right. I got it done

and changed course to the north, over the land. Visibility was incredibly bad. There was a heavy brown haze in the air and at my cruising altitude of 9500 feet I lost contact with the ground. Once again I needed radar guidance to land at Bombay airport. I didn't see the runway until I was 200 feet away from it.

13: BOMBAY TO SINGAPORE

Fortunately for me QANTAS maintains a presence in Bombay, and Tommy Tiemar, the QANTAS man there for over twenty years, met me and helped me through the interminable complexities of the Indian way of doing things. It still took three hours, but the people were pleasant, and some of the officials were women. It made all the difference.

By the time I got to the terminal entrance and organised a cab to my hotel I felt as worn out as I had following my epic nights over the Pacific and Atlantic Oceans. It was nearly thirteen hours since my arrival at Karachi airport that morning. That day I'd spent more than twice the time going through airport absurdity than I had in flying. What a world.

The hotel was beautiful and I enjoyed a wonderful Indian meal before enjoying a deep, relaxing sleep.

I answered a knock on my door the next morning and was greeted by two beautiful Indian women, Mohini Schroff and Saudamini Deshmukh, both 99s and the first 99s I'd had contact with since leaving Florida. It was so great to see them. Saudamini has the distinction of being India's first commercial woman pilot and is a captain with Indian Airlines, the domestic carrier. At last I was getting back to a world that was more familiar, a world where women are able to break down the barriers and seek their role in life on an equal footing with men. Give us women 'a level playing field' as politicians and businessmen like to say, and we'll show the male section of society that we can play the game as well as they can. Of course, there's one other proviso — make sure to give us the same rules, too.

Mohini and Saudamini organised a press conference and the Indian press was very interested in my trip. One of the press members noticed that I wore a locket with a photo of Sathay Sai Baba. Sai Baba is a modern day saint, I feel sure. There have been so many recorded eyewitness accounts of the miracles he has performed that there is no doubt in my mind. A friend of mine in Sydney had given me the locket to bring me good luck on my journey. I also had a St Christopher medal, a rabbit's foot my children had given me, and the piece of fabric from the Southern Cross given into my keeping by Dick Smith. No wonder I felt protected.

I was not a devotee of Sai Baba, but I had read enough about him to believe that he was consciously in touch with a greater power. I was more than happy to carry his photo and, often, during my most harrowing moments, his image would pop into my mind.

The fact that a Westerner would carry his photo really intrigued the Indian press, but I was very aware that India has a strong spiritual history which many Westerners have difficulty coming to terms with, especially when they look at the obvious poverty that exists in India. However, the mystical qualities of India and the eastern way of spiritual understanding requires the putting aside of the rigid belief systems that have confined Western religions.

I went shopping with Mohini in the afternoon. We took a pedi-cab and our driver tore his way through traffic that was so chaotic I was unable to fathom how people got to where they wanted. We got our shopping done and got back to the hotel, thanks to the kamikaze attitude of our driver.

I was invited to dinner that night with the Australian Consul-General, Gavan Bromilow, and that morning I had phoned him and asked if I could bring along Mohini and Saudamini. He reluctantly agreed. I'm not too sure that he understood my motives. I was selfish. It was so long since I had enjoyed the company of women that I wanted to keep them around, and they wanted to spend an evening with me. The Consul General was gracious about it and we enjoyed a delicious meal along with two officials of the Australian Government who were staying at the consulate.

Tommy came to the hotel early the next morning to take me to the airport and help me through the maze of airport officialdom. Despite his assistance it took me four hours to get through everything and get to my plane. Tommy and I took turns calming each other down. To see an Indian become frustrated at the delays and circumlocutions of Indian bureaucrats was a heartening and humorous sight. The crazy thing about it was that I was only making an internal flight to Madras on the east coast.

It was hot and the airport was smog bound. I was kept waiting in my plane for 20 minutes. I though I would faint from the heat. My auxiliary fuel boost pump didn't work and I prayed my plane would start without it. The engine ticked over first go. I got to the point of take off and airport control directed me to maintain runway heading and report again at 2000 feet. This was a normal request. I took off in the heavy smog conditions and at 2000 feet prepared to report in as ordered, when I noticed a tree branch appearing out of the smog almost brushing my left wing.

Oh, shit, we're in trouble.

Stay cool, Gaby, whatever you do stay cool.

Very easy to say, I thought, as I pulled the nose of the plane up and climbed past a chimney stack off to my right. I put my plane in the best angle of climb possible. The idiots at Bombay control had put me right in the middle of a hill.

At 9500 feet I relaxed, when the controllers came back on and ordered me to fly at 12,500 feet, despite the approved 9500. 'God, can't I get these dick heads to understand that I'm flying a small plane without oxygen and not a jet.'

No, Gaby.

Ha ha, very funneee.

But no amount of asking would get them to change their

minds. I was stuck at 12,500 feet for my four and a half hour flight across India. I soon ran into heavy cloud, rain and turbulence because of the monsoonal conditions to the south and with only one NDB on my track to Madras I hoped I would maintain track. The navigation aids were not particularly reliable.

I tried many times to raise Madras using HF radio. I tried again and raised the Cocos Islands in the Indian Ocean off the North-West coast of Australia, but no luck at all with Madras. I gave up. I also heard a QANTAS plane overhead, but that was my only contact with the outside world. Once again I was alone in my capsule tossed around the sky by angry weather.

I was sweating on picking up the NDB at Madras. I picked it up right on target. I was still in cloud but at least I knew I was on target, and raised Madras tower on VHF radio. An Australian helicopter pilot was in the area — his accent was a giveaway. He was only 20 miles out of Madras and inbound. He told me he was in clear skies and would wait for me on the ground. What a gentleman, I thought. We spoke on the radio for some time. His name was Peter Mankleton and he was servicing oil rigs in the Bay of Bengal. He arranged an engineer to check out my fuel auxiliary pump.

I broke into clear weather about 25 miles out. Oh boy, I thought, everything's great.

Peter was waiting at the parking bay and on his advice, and following my previous experience in Bombay, I went to the Flight Service office to make sure all my departure clearances were okay. I especially wanted to make sure I was cleared to fly over the Andaman Islands in the Bay of Bengal, which were on a direct line to my next port of call, the island of Phuket off the Thai coast. The Indian Government has a Naval base on the main island at Port Blair and air space around the islands is restricted. Tommy in Bombay assured me that clearance was organised, but I needed to make sure. Thankfully I did.

The people in this part of the country are Tamils and very excitable. The Flight Service personnel were no exception and I thought I was watching a Peter Sellers parody as they emphatically told me it was impossible for me to have such a clearance.

Here we go again, Gaby.

No, I'm buggered if I'm going to put up with this garbage again.

Don't get angry, Gab, it doesn't always work. Try being agreeable for a change.

Very funny. Okay, I'll be agreeable.

For some reason I took my own advice and decided to agree with the Flight Service personnel. I did a complete about face and told them they were probably right (in fact they were and, like so many people, I'm sure Tommy gave me false information to make me feel good at the time). I then proceeded to give them a lecture about the dangers of the rules and regulations of Indian bureaucracy and how I found India to be full of lovely people except they were being strangled in their expression by all the red tape.

The alternative to flying over Port Blair was to travel south-east for many miles to a point south of Nicobar Island and then turn back north-east to Phuket. I told them that this was unacceptable to me as I was flying a single-engined plane and had no way to fix my position and I needed the NDB at Port Blair to check my position. I also said I believed in the power of the individual to move mountains and cited Ghandi as a great example of this.

'I will come back with the correct approval if only to demonstrate to you that one person can overcome the power of the Indian bureaucratic machine,' I told them as I dramatically exited their office. I think I laid it on a bit thick, and was probably suffering from brain fatigue caused by all the nonsense.

I had no idea how I would get the approval, but something told me not to worry. Maybe I could just take off and fly that route anyway. Peter talked me out of that thought with the story of two Americans who'd deviated only slightly from their flight path over the mainland and had been intercepted by a jet fighter and forced to land. They spent two days in jail and a month in custody until the authorities allowed them to leave.

I stayed in the Hotel Trident, which is a very modern hotel with excellent service and a menu featuring exquisite Indian and Thai foods. A friend of Saudamini, named Sanita (Sonu) Shahaney, arranged to take me shopping the next day. She called around to collect me in one of those Indian-made cars which look so much like the cars of the '50s. Sonu was a strikingly beautiful woman with a creamy white complexion and black satin hair. She asked me what I wanted to buy and told the chauffeur where to take us. We went back to Sonu's house for afternoon tea and when she heard about my troubles in gaining clearance for Port Blair she told me she would help me and phoned her husband Ram, who is the General Manager of one of the largest truck and bus builders in India. After the phone call she told me not to worry, that Ram knew someone who could organise it, though it might take some time.

During our time shopping I had told Sonu that I was interested in Sai Baba. This was great news to her as Ram's sister Leila, who lived with them, was a devotee who had been to see him several times. It looked like I would have to stay over the weekend to get the important clearance, so Sonu went ahead and organised for Leila and me to catch a flight to Bangalore and go and visit Sai Baba's ashram, Whitefields, about an hour's drive from Bangalore. Before I knew where I was, I was on the plane to Bangalore.

The plane was an Indian Airlines 727 and I couldn't believe the condition it was in. The interior was dirty, with scraps on the seat and floors and marks all over. I thought about the immaculate condition in which Australia's domestic carriers keep their planes as I struggled my way into a narrow, stained seat. I wondered what the maintenance was like. I knew that the Indian way of doing things was different from the way I was used to and this wasn't necessarily wrong or bad, just different.

I was invited into the cockpit and spent time with the captain and first officer. Boy, what a job they had. They were flying a passenger-carrying commercial jet without weather radar, without a transponder and without much in the way of navigation aids. They only had one ADF and two VORs. They also told me they did six flights a day, six days a week and the salary package was very poor.

We arrived in Bangalore and were met by a car and driver organised by Ram. We spent that night with a friend of Leila who lived in a cottage built for the British Raj. It even had the crown embossed on a parapet in the front.

Early the next morning the driver arrived and we set off expectantly to see Sai Baba. We both became silent and reflective in the car and I began to review my life. All sorts of forgotten facts came up. I remembered in detail a china figurine I'd had as a kid. It was like the rewind of an old movie. I saw my mother and the courage she'd shown when, after Laurie's death, she'd received a small inheritance and, borrowing money, bought a business. This was in 1960 and very few women at that time were able to borrow from banks. I also saw again the courage she'd shown when facing her death.

My thoughts moved on to what I might ask Sai Baba if I was granted the privilege of receiving a personal interview with him. I pondered many questions I could put to him, but none of them really seemed to matter. Maybe I could ask if I'm on line with my destiny, I thought.

Gaby, you're really doing fine, aren't you?

Yes, yes I am. I don't believe I need to ask him anything. Then the realisation hit me that I didn't need to ask him anything — all of the answers were inside me. Perhaps I'd come all this way to find that out. If so, it was well worth it.

The ashram was surrounded by a high pink wall and as Leila and I entered I was overcome by a deep feeling of inner peace. I couldn't recall having that depth of feeling before.

Sai Baba was due to appear among his devotees in a circular building with open sides and a roof supported by columns. There was a raised dais in the centre and an ornate chair for Sai Baba. Chairs were placed around the dais in

a circular pattern. We managed to gain seating near the front
and I was fascinated to watch the obvious care and loving
with which his devotees laid garlands of flowers over his
chair and spread petals on the dais and walkway to it.

Soon a group of musicians struck up and the large
gathering started singing bajhans. Bajhans are sacred songs,
usually with a simple melody and verse structure which is
repeated many times. They have the effect of lifting the
vibrations or energy levels in much the same way group
chanting or Gregorian chanting does. The bajhans sung that
day were written by Sai Baba. It was the first time I'd had
this experience and I felt myself crying. There was no real
reason for it, but tears of release poured down my cheeks.

This went on for about 15 minutes, then the tempo
suddenly increased, which indicated the imminent arrival of
Sai Baba, and, as he approached, the music became quite
frenetic.

He appeared, walking barefoot, hands together in front
of his chest, in a saffron robe. He walked over the petal strewn
path. As he passed the row in which Leila and I were sitting
he turned and with a nod of his head, recognised each of
us in turn. I looked into his eyes and experienced the greatest
feeling of love I have ever experienced, and with it came
a sense of peace, a knowing inside of me that everything
was all right — with me and with this crazy world of ours.
It was truly the peace that passeth understanding.

Since leaving the Catholic church I have never followed
a set religion. I went though a period of atheism, but came
to the understanding that the thing we know as God is really
not definable, except as absolute love. I'm sure that's the
essence of Christ's teachings and the essence of every great
religion and something in that look from Sai Baba said: 'That's
right, Gaby.'

He stayed with the group for about ten minutes and
sang bajhans with us and then walked among the crowd,
giving his blessing to many people. Some of the people there
threw themselves at him, to touch his robe or his feet. They
were like teeny boppers throwing themselves at a rock star.

I thought they were putting too much on him, instead of using his wisdom to look within themselves and find the spark for themselves.

He then left with the announcement that there were to be no private sessions that day, but I knew I didn't need one. I had already received what I came to get.

That evening Leila and I flew out of Bangalore and arrived back in Madras about 11.00 pm. Sonu and Ram met us at the airport and dropped me off at the hotel. I was in a very contemplative mood, which lasted through the Sunday.

On Monday I decided to go to the airport and make sure I could do everything possible to hasten my departure the next day. I was convinced that the clearance to fly over Port Blair would come through, as Ram had told me he would push it along. I wanted to get an early start the next morning, as I expected the flight across the Bay of Bengal to take about 9 hours.

I spent most of the day at the airport, lining up the departure clearances, and at 4.00 am I got word from Ram that the clearance had been granted and so, armed with the clearance number, I marched into Flight Service to submit my flight plan over the Andaman Islands.

'Oh no, no, no. It is not possible'. The three Indian men started to get excited. I waved the number in front of them. They thought it was fiction.

'Okay', I demanded, 'take me to your boss'.

I was ushered into the inner sanctum to receive the same denials. I let go. 'Get on the phone,' I ordered and gave him the name and phone number of the senior official in Delhi who granted the clearance.

They tried ten phones before they found one that worked. By this time it was 5.15 pm and I prayed that he was still in his office. At last the connection was made and at last they believed I had the necessary clearance. My perseverance had paid off, along with my determination to show the men in Flight Service that the grinding wheels of their own country's bureaucracy could be halted, although I think it went over their heads.

I went back to the hotel knowing I had done everything possible to hasten my way through the airport's officialdom the next morning. Oh, well, I guess it did pay off. It only took me two hours from the front door to my plane. Sonu, Leila and some friends showed up to wave me off and one of the women gave me a lovely soapstone carving of the elephant god.

I took off, a little depressed at the delays, but so very pleased to be getting out of India. Once again I had been in a country where the people were mostly helpful and generous, except when they were working in an official capacity and had a paper to stamp or a form to fill in. I contemplated the obvious poverty in India. I saw so many examples of endemic poverty, but I also saw a people who were happy. In other countries I'd been in where poverty was widespread, the people looked sad and totally without motivation. However, in India the people were smiling and active, and I saw their acceptance as a two-edged sword which gave them a wonderful forbearance, but also held them back.

My flight across the Bay of Bengal was mainly spent avoiding storm cells. I used the NDB at Blair Point to good effect and as I tracked across the Andaman Islands I was able to see glimpses of jungle through the clouds. I contacted Port Blair tower in a cheerful voice, but the controller on duty gave me a curt reply. He was probably disappointed that he couldn't scramble a couple of jet fighters to intercept me.

I approached Phuket airport in the late afternoon and made a copy book landing. Everything's going fine, I thought. I should have kept my mouth shut. Two ground marshalls guided me into a parking bay and just as I got the okay signal from them I heard a *prrrrrt* noise from the front and a vibration ran through the plane. Oh damn, something has hit the propellor, I thought. I shut down, and as I was scrambling out of the plane I noticed one of the marshalls pull a concrete block away from the front.

I jumped out as they started walking off. Sure enough, my propeller was bent. I turned on those two and yelled out at them that they'd bent my prop. 'No no', they assured me, 'it must have been like that before'.

They were laughable. I knew I was not going to be leaving Phuket the next day, so I called on the forbearance I'd seen in India and decided that, as night was falling, I would have to see to it the next morning.

Getting through the Thai officials was a dream and I was through all the formalities in five minutes, wishing the Indians would model their clearance methods on Thailand. My hotel was the Pearl Village, a beautiful place with pleasant, friendly staff, who unfortunately didn't speak English very well. The next morning I phoned Dowty Aviation in Singapore, because they were due to service my plane when I arrived there. I wanted to know if my plane was flyable. They didn't think so, but told me if I could fax them a tracing of the bent sections of the propellor they could give a more definitive answer.

I also phoned Sydney and spoke to Sherry Stumm and my old friend Pat Wilson. Pat had organised my plane's insurance through G.M. Forsaith, the Heath Fielding Group. She told me she'd contact them and get back to me. I went to the airport and took the tracing and faxed it to Dowty Aviation. I also called in at the airport manager's office and reported the incident to him.

When I got back to the hotel there was an urgent message from the insurance company telling me under no circumstances to fly the plane and that they were sending Richard Ong, an assessor, from Singapore. I later heard from Dowty Aviation that they considered the propellor too damaged and, following Richard's report, they were able to find a new propellor in Sydney, which was flown by commercial flight to Singapore and then to Phuket. Dowty Aviation sent an engineer up to carry out the work.

All this took time — six days in fact. I was so close

to home that the lazy days by the pool at the Pearl Village threatened to become a weight around my neck. I tried to think about all the situations during my flight when things hadn't gone my way and, how I'd resisted and pushed trying to make things happen the way I wanted, until I realised that my lesson was to let go, accept and relax. Each time I'd done that the situation had turned around and I eventually got what I wanted, though not always in the way I expected.

So, I decided to enjoy the relaxation the Pearl Village offered and ended up with severe sunburn. I also went exploring one day with two Aussie girls in a four wheel drive and, although Phuket is a small island, we almost got lost in the jungle, but we finally found our way back to the hotel, exhausted but satisfied with our adventure.

At long last the time came when all was fixed and I could get on with my next leg down the west coast of Malaysia to Singapore. Conditions were overcast and this was the monsoon season, but I was fortunate and had relatively good flying conditions. I had left Amelia Earhart's flight path in Karachi. From there she had flown directly across India to Calcutta and from there to Rangoon in Burma, a place that was unavailable to me for political reasons. From Rangoon she'd flown to Singapore and then to Bandoeng Island in Indonesia and from there to Darwin, or Fort Darwin as it was known in those days.

I was concerned about locating the right airport in Singapore. I was due to land at Seletar, a joint civil/military airport with all the required entry facilities, which was mainly used by light planes. Singapore is a small island and there are several airports dotted around it. I thought I might have difficulty locating Seletar as it was raining and ground features were hard to make out, but the controllers had me on radar and I was vectored in to Seletar.

I parked near the Aero Club of Singapore and was greeted by the local media, who peppered me with questions. Also there, to my delight, was Marcel from my sponsors, Slender You. He presented me with a large koala bear. I was really getting closer to home.

Dowty Aviation was nearby and I checked in my plane with them, happy that it would be in their professional hands. My stay in Singapore was to last three days because of the plane service.

Though I had been in Singapore before, I was still impressed with the productive and innovative dedication shown by these hard-working people, who have raised their tiny country to be the true pearl of today's Orient. I saw building construction sites where building workers laboured away in three shifts around the clock. The good old protestant work ethic is alive and well in Singapore.

I enjoyed a memorable meal with Marcel and Gerard, a friend of Marcel, who was in the Singapore advertising industry. He used his local knowledge to take us to a delightful restaurant, where we ordered a dish called drunken prawns as the entree and chilli crabs as the main dish. The drunken prawns were cooked at the table. I think the name of the dish is enough description. Marcel was leaving for Sydney the next day, so we toasted his departure and I told him how much I was looking forward to being with my children.

When I went back to Dowty Aviation to collect my plane I was asked where I got the snake. *What snake?* They told me how, when one of their service men was vacuuming the back section of the cabin, a snake shot out from under the auxiliary tanks, slithered down his uniform, onto the hangar floor and headed down a drain pipe before anyone could gather their wits.

I couldn't believe what I was hearing and told them to stop pulling my leg. They went and got the man. He swore it was true and told me that the snake was green and yellow and about a metre long. Apart from that he said he didn't have the time or inclination to identify it.

I was dumbfounded. The only place I could think where I had parked on grass was at Rabaul, almost a full circle of the globe away. If that snake got on there it must have had a hair — oops! — scale-raising time of it. I could see how it would have made a comfortable home for itself tucked under the auxiliary tanks where no one could possibly interfere with it. What it ate on the way I don't know.

Thank God it didn't decide to crawl out while I was flying. Perhaps it did, but at least it didn't disturb my space.

When it came time to leave Singapore, Gerard saw me off at the airport. The rain was pelting down, but I didn't care. One more stop at Bali and I would be in my own country again.

14: SINGAPORE TO SYDNEY

Storm clouds were marching across a threatening sky, but by now I was adept at dodging them.

I was in the turbulence of the Inter Tropic Convergence Zone again and due to cross the equator for the fourth and last time, just south of Singapore. I couldn't afford to become complacent, certainly not this close to home. I was going to make sure I made it. If I had been on my original schedule I would have missed the monsoonal influence, but I also would have missed many of the adventures.

The Australian media was right behind my flight and there was a great deal of excitement building up towards my arrival, which I was dreading. I didn't think I'd be able to handle it all, after so long alone. *There you go again Gaby, setting yourself up for failure. All you've got to do is be there and be yourself. The people aren't expecting a great orator to appear amongst them, and besides if you do make a boo-boo people will know you're human and be with you more because of it. Anyway, you're days away from it, so don't worry and just concentrate on getting us home.*

Once again the advice was right.

So I crossed the equator and made my way through the islands of Indonesia, picking up strange snippets of radio conversations between pilots. The only incident which marred this leg was an Indonesian controller who mocked my voice when I reported my position and ETA for Bali.

All the words used to describe Bali are true. It certainly is a jewel in the Java Sea. I detoured around the majestic peak of a volcano reaching 4000 metres into the clouds and saw the island set out below me, dazzling me with the richness of its colours.

I was met at Denpasar airport by very helpful airport staff. Bali is a favourite destination for young Australian tourists and QANTAS has a strong presence there. My hotel was close to the beach and I was shown to my Balinese bungalow by the friendly local staff. That evening James Elliott, the Australian Consul and good friend, joined me for dinner, along with Joel Strickland from QANTAS, and after the meal we went to a local market, where I bought some Balinese kites for James. I went to bed that night pleasantly tired and so looking forward to tomorrow — the day I would touchdown in Australia again.

I woke to the early morning sun streaming through my window and the sound of the waves lapping on the nearby beach. This was it, I told myself. This was the day. The day of my last water crossing. The day I would see my own country again. The day I would touchdown at Darwin, Amelia Earhart's only stopping place on the Australian continent and the place where, I felt I could say goodbye to her. The tingles of excitement began to build up.

The airport staff picked me up at the hotel early and, with their help, I cleared the airport officials in record time and headed out into the Timor Sea longing to see land.

My first contact with Darwin came over the long range High Frequency radio, and though it was good to hear an Aussie voice again, the operator's was very official, and a bit of a let down. Fortunately the situation was saved by the operator in Perth, far to the south, who'd picked up my call and came on the air to say the words I was longing to hear. Three simple words, packed with so much meaning: 'Welcome home, Gaby!!'

Well I wasn't home yet. Still 500 miles to go, but it was

certainly getting closer. I looked down at the sea below me
and the heebie-jeebies about the water and sharks started
to get to me again.

There was a big thunderstorm over Darwin and I hoped
it would blow away before I got there. This was early
November and the beginning of 'The Wet', that season of
torrential rain and fierce thunderstorms that turns much of
the top end of Australia into a series of water ways teeming
with life. I made contact with Darwin Flight Service over
the shorter range VHF radio. I was only fifty miles out and
beginning to curse the cloud that obscured the coast I was
so looking forward to seeing.

An enterprising local press reporter used her initiative
and conducted an interview with me over the radio. 'What
are you looking forward to when you first land?' she asked.

'A nice cool beer,' I replied.

Twenty miles out the Darwin Air Traffic Controller read
out to me the following message from the Northern Territory
branch of the Australian Women Pilots Association (AWPA):
'Gaby, welcome home. We are truly proud of your magnificent
efforts and applaud you as you join other aviation legends
like Amy Johnson, Jean Batten, Amelia Earhart, Ross and Keith
Smith, Bert Hinkler and Charles Kingsford-Smith.' This was
a special moment in my journey and I was overwhelmed with
emotion.

Ten miles out I saw the coast of northern Australia. My
excitement increased. Three Cessnas from the local Aero
Club came out to meet me and escort me in. At last we
crossed the coast and did a circuit of the city which had
been devastated by Cyclone Tracy in 1975 and now lay below
me like a king size welcome mat.

I landed on the main runway and, as I slowly reduced
speed, I saw two fire trucks, one on each side of the runway.
Gosh they must have expected me to crash, I thought, but
as I approached them, they turned on their water cannons
and I taxied under a welcoming arch of water. I instinctively
ducked and closed my eyes. As I taxied to the parking bay,
I saw a crowd of people gathered. My first thought was that

they were there to see off an international flight. I knew my sister Sheri was going to be in Darwin to meet me, but I didn't expect that the people of Australia's most northern city would come out to welcome me home. I was so moved and felt my heart pounding away.

Take a deep breath Gaby.

I stopped the plane and an immigration official knocked on my storm window and passed me through the inevitable can of insecticide which I sprayed inside my plane. Many visitors to Australia are puzzled or annoyed because the interior of every aircraft is sprayed before passengers are allowed to disembark. It might seem a strange custom, but it has kept many exotic pests and diseases of humans, animals and plants out of Australia. I was happy to do it; after all, it is an Aussie custom.

As I opened the door a brass band struck up *Waltzing Matilda* and the crowd broke into cheering. I was overcome and had to fight to keep back the tears. I couldn't believe that all these people were there to see me. It was overwhelming.

Sheri was there with a representative of the mayor and the local Member of Parliament and we hugged and hugged and wept on each others' shoulders. It was so good to see her. My little sister. She was so proud of me and pressed something into my hand. It was a poem she wrote for me.

To Gaby,

> *She wandered through the mist of*
> *thoughts, provocative and awe inspiring.*
> *For her spirit, her core, her inner peace*
> *was still unmet.*
> *Those thoughts reached a plateau as*
> *yet untouched.*
> *A whiteness, a truth, an understanding.*
> *And she began the lonely trek through*
> *adversity and oft disdain.*
> *Until for her the calm of soul commences when*

She recognised,
She accepted,
And began the challenge
She met.

With love and admiration
Sher.

My beloved sister Sher. We had been through so much together as children and here we were in Darwin together again and reunited by the bond of love we will always share.

The rest of the day went by in a haze. I know I kissed the ground at Darwin airport. I wanted to acknowledge my country. This country, in which I was so fortunate to have been born. I know the Pope kisses the ground at all the airports in which he lands. It is his gesture of acknowledgement to the people of that country. I wanted to acknowledge the country itself. The actual earth. Unfortunately, I kissed 15 centimetres of hot concrete baking in the tropic sun. Somehow or other it didn't feel right for me, and I had sympathy for the Pope.

It seemed there were many people there that wanted my attention, but I was happy to walk through the crowd, signing autographs, accepting small gifts. Two people must have known of the press interview and pressed cold cans of beer into my hands. The people of Darwin were showing me the same generosity they'd shown Amelia Earhart when, in 1937, it was Fort Darwin, a small tropical outpost thousands of miles away from anywhere else in Australia.

Australian flags were pressed into my hands and I posed for many photographs. It was so good to be amongst Aussies again and hear those Aussie accents that told me I was home in my own country. The officers from the Federal Airports Commission were so thoughtful and had arranged space for a reception and media interviews. It was a hectic time and after the euphoria wore off I found myself getting tired. I think everyone sensed this and Sheri and I were whisked of to the Sheraton-Wentworth Hotel where we shared a suite decked out with flowers provided by the management.

It was so good to put my feet up and to share some quiet time with my beloved sister. My plane was under the care of the Royal Australian Air Force and I was content not to worry about it for two days while I fulfilled a lot of engagements around Darwin. Kathryn Flynn, a 99 and Australian Woman Pilot, had arranged a reception for me and a barbecue at her home with other women pilots. I visited schools and spoke to the children, gave many media interviews and added my signature to the wall of the old Victoria Hotel, which bore the signatures of many famous aviators and pioneer aviators. The historic hotel, which survived the ravages of Cyclone Tracy, is now incorporated into a shopping mall.

Kathryn also organised a dinner reception in the Sheraton-Wentworth ballroom, which was decorated in the most exquisite way, with shades of blue everywhere. I was asked to make a speech to the several hundred people there. It was my first public speech. I was very nervous and started off in a halting manner and then thought, what the heck, I might as well just let go and be myself. I'm not too sure what I said, but the applause told me I hadn't made a complete fool of myself.

I told the people of Darwin that it was my intention to raise $1million for the purchase of a new plane for the Royal Flying Doctor Service, which does such a magnificent job in providing medical attention to the scattered peoples of the outback. People presented me with many gifts, including books and a special painting, which now hangs in my office.

After two days of engagements it was time for me to go on to my next stop in Alice Springs. I was going to change my route from there to fit in with some new arrangements that were going on in Sydney. Because of the interest in my arrival it was thought that it would be a good idea for me to fly to Parkes in western New South Wales and for my children to be flown there, too. The arrangements were kept hush-hush to give us the chance to spend some time together before all the attention in Sydney.

Sheri came to the airport to see me off. She was due to leave at the same time and the captain of her Ansett Airlines flight, Captain Hanrahan, very courteously held up the passenger flight so we could say goodbye. I took off into clear sunny Australian skies only minutes before Sheri's flight and I was still climbing to cruising altitude when I heard Captain Hanrahan call me over the radio. I answered, and in a gallant gesture he passed the radio over to Sheri. We spoke briefly and said our goodbyes again.

I was alone in the skies once more, but these were familiar skies. The outback stretched below me and met the horizon in a haze where the land merged with the sky. How great it was to be flying over this land. The spirit of Australia filled me with its warm greeting. I tracked to Alice Springs over the impressive new Air Force base at Katherine and over the immense stretches of the cattle stations and Aboriginal lands of the Northern Territory. No wonder they know this land as sacred. The Air Traffic Controllers were friendly and professional. They are one of the reasons flying in Australia is a pleasure.

I was met at Alice Springs airport by members of Zonta International, The Australian Women Pilots Association, the Royal Flying Doctor Service, as well as the local television reporter, Irwin Chandler, and his wife Kieran. I have known them for a long time and was pleasantly surprised to see them. I was able to park my plane at the Royal Flying Doctor Service hangar.

I stayed at the Sheraton-Wentworth. Tom Selleck was staying there. This is great, I thought, I'll be able to tell Mimi I saw Tom Selleck. And I was given a reception by mayor Leslie Oldfield and Mary Blacklock of Zonta at the Council Chambers and, boosted by my accomplishment in Darwin, gave a little talk. Dinner that night was in the hotel with a group of women pilots. We had a fun night. Marcia, who met me at the airport, flies a helicopter out of Alice Springs and is a delightful woman with a strong sense of humour. She did my flight plan for me next morning.

I headed for the airport early the next day, pleased that

all I had to do was file the flight plan and check the weather. No more customs, no more immigration, no more facing up to angry looking men carrying weapons and doing their best to show me the petty power and authority they had. No more, no more, no more. I was truly home.

I took off at 8.00 am, heading south-west across the north-east corner of South Australia and into New South Wales. About an hour out of Alice Springs I tuned my radio to the frequency used by the Royal Flying Doctor Service School of the Air and for about half an hour I answered questions from the kids in those outlying communities. They wanted to know a lot about my trip and I'm afraid the half hour wasn't long enough. I was pleased to be able to do it for them.

I think we Australians have a lot to be proud of. Our country is about the same landmass as the continental area of the USA and yet our population is about the same as that of the greater Los Angeles area and we have the facilities available to us that many peoples of the world don't have.

Flying over the sparse interior landscapes reminded me again of the forgiving nature of this land. We are blessed indeed.

I crossed over the Darling River and was not far out of Parkes when I was met in the air by Jennifer Byrne from *60 Minutes* in a chartered plane. I had last seen her in Atchison and we had a good talk. Jennifer and the *60 Minutes* crew organised to do an interview with me in the motel in Parkes later that afternoon.

I landed and parked at the Aero Club, then went in and had a beer and sandwich with my friend George, the Chief Flying Instructor, and his wife Ingrid. George and I studied navigation together at Sydney Technical College.

I was starting to get excited at the thought of my kids arriving tomorrow on a commuter flight. Neville also suggested that he might fly down that afternoon and I wondered if he would come and how I would feel about it if he did.

We conducted the *60 Minutes* interview in my room

and it went for a long time. Towards the end of it, just when Jennifer was asking me about the future prospects of Neville and I getting together, the phone rang. I knew it was Neville and thought he might be phoning from Sydney to say he wasn't coming down. It was him all right, but he was phoning from Parkes airport. I started to feel butterflies. The *60 Minutes* crew were very understanding and fortunately we were just about at the end of the interview.

Neville arrived at the motel and it was nice being with each other. I knew I didn't want to go back to the kind of life we had previously led and I now felt very independent. My journey around the world had shown me I could accomplish the things I wanted to do with my life and gave me the strength to tell Neville the things that I wanted from any future relationship. We were both pleased that the future held good possibilities for our relationship, but it had to be on a partnership basis with both of us taking responsibility for making it work.

Neville flew out the next morning before the kids arrived. He knew I wanted my own space with them and I respected him for giving me that space. He had seen a lot of them during my absence and had sometimes stayed over.

At last their plane appeared and, before I knew it, we were in each others arms and hugging and kissing and jumping up and down with the joy of seeing each other. Oh gosh, they've both grown so much, I thought. Mimi looks so mature and I'm sure James has grown at least two centimetres. They wanted to know about the delays that kept me from them for so long. My plan had been to be away six weeks — 42 days. Today was day 98, well over twice as long, and by tomorrow I would have covered 29,000 nautical miles and spent 240 hours in the air.

We spent the morning in the park, on the swings, feeding the ducks and just generally being in each other's company. It was so great to be a Mum again with her children. My capacity to love and appreciate them was so much greater.

That afternoon we talked in the motel while I prepared

myself for the morning's flight into Sydney. I saw the kids off and returned to the motel for the last night of my flight.

I don't think I slept much at all that night and was up early the next morning. Day 99, the last day of my flight. I wondered if there was a coincidence that it should be the 99th day, the same number as the 99s, the group founded by Amelia Earhart.

The Royal Aero Club at Bankstown was sending a group of formation flyers to escort me into Sydney. There had been a ten-page fax explaining the details of the formation and arrangements to meet at a small airstrip at Camden for a briefing of our planned Sydney flyover.

I took off in clear skies, though the weather report for Sydney was for overcast conditions with some slight drizzle. I made the rendezvous point at Camden a little late and was briefed on the formation flight to Hornsby, along the northern beaches and through the heads to fly up Sydney's magnificent harbour, holding over the bridge and the Opera House, past the central business district and over the suburbs to Bankstown and the final touchdown of my flight.

Everything worked like clockwork. It was overwhelming. The sky over the harbour was busy with helicopters and light planes. I had to concentrate on my flying, there was so much traffic in the air. Media helicopters buzzed in close to get live television coverage of me talking to their reporters. Dick Smith was nearby in his helicopter and asked me about the fabric from Smithy's Southern Cross. It was safely pinned to the inside of my aircraft. What a great guy he is. I think Dick took just as much pride in me successfully completing my flight as he did in his own many successes.

It was a moving experience for me to fly over Sydney and look down at the familiar spread of this beautiful city. I saw North Sydney and spotted my own house. It was moving to know there were people looking up and wishing me well. The emotion of the moment started to get to me and I found myself breathing hard and deep to gain the control I needed to safely pilot my beautiful little plane to the last landing on this long, long journey.

Bankstown control had issued my name as the code name of the day for airport conditions and the controllers seemed just as excited as I was as I gently lowered my plane over the rooftops and onto runway 29er right. If I thought the reception at Darwin was overwhelming, I don't have the words to describe the scene at Bankstown airport as I leaned the mixture and my engine turned on its last revolution as the propeller came to rest.

I remember seeing black and white films of Sir Charles Kingsford-Smith, Charles Ulm and the great pioneer aviators of Australia being swamped by people as they landed their planes after a record-setting flight from Europe. I didn't ever think that something similar would happen to me. There were so many people there. As I crawled out of that cramped cabin for the last time, Howard Knox, the Bankstown airport manager, placed a laurel garland around my shoulders. I remember people crying out to give me room, but the people of Sydney were so keen to share my moment of triumph with me, that I almost found it impossible to move. It was exciting to receive congratulations and good wishes from so many people.

Eventually I was able to make my way to a dais, where I was greeted by Kay Cottee, Bronwyn Bishop, Dawn Fraser and Cliff Young and received a welcoming address from Catherine Greiner, the wife of the New South Wales Premier, Nick Greiner. I'm not too sure what it was I replied. But I do know that I said I was pleased to be home, and I meant every word of it.

I hardly had time to catch my breath before I was whisked off to the city for the motorcade up George Street, from Circular Quay to the Town Hall. Mimi, James and I were seated in an open car. It was okay at first, because I was still in a daze and there weren't so many people about at that end of town, but then as we got closer to the Town Hall there were more and more people.

We passed Martin Plaza and the crowds along the footpath grew deeper. People were waving and cheering. 'Good on you, Gaby! Great work, Gaby!' Cries of congratulations and cheers filled my ears. I had to pinch myself. I looked around at my children. They were here with me. This really was all happening.

Tears began to well up in my eyes. Oh gosh, help me. I can't cry now. Not now, not now. *You don't need me now, girl, you've done it. Your here. My job's over. I just had to help you accomplish your dream. Congratulations, Gaby, and remember ... keep dreaming.*

As we neared the Town Hall a brass band was playing and the noise of the crowd was unbelievable. There were many women in the crowd. I could see the obvious pleasure on their faces. I hadn't only done this journey for me. I had done it for them, and all those women who feel they are caught in a rut at home with their children, or those single mothers working to raise their kids, or the men who feel they are caught in a vicious cycle of work/mortgage/work. I had done it for all of us who can dream.

My car pulled up near the Town Hall steps and a senior ranking police officer placed the laurel garland around my shoulders as a piper escorted me up the Town Hall steps, where microphones were set up. Television news crews, radio reporters and press photographers were jostling with the people. I knew these people were there for me, and I also knew that I was there for them.

Sydney's Lord Mayor, Alderman Jeremy Bingham, came to the microphone and presented me with the key to the city of Sydney. 'This is not just the key to our city,' he said, 'This is also the key to our hearts.' Only twelve other individual people in the history of Sydney had been given this honour. I was the thirteenth. The lucky thirteenth. I felt very humble to be honoured in such a way by the people of my home city.

I found it difficult to deal with the emotional impact. But I had done it and I knew I could reply to Alderman Bingham's address. I remembered that Shana Bellin had excitedly whisepred to me on the dais at Bankstown, 'They brought the Berlin Wall down today.' And as I thanked Alderman Bingham, the Council and the people of Sydney for the honour they had done me, I thought of the many barriers I had encountered and broken down during my world flight, and hoped many other people could do the same in their own way.

EPILOGUE

So, that was my flight around the world.

I had a dream that I wanted to make a reality. There didn't seem to be much purpose in my life until I discovered that dream. Neville had left me. I was a single mother with two kids and I hadn't really accomplished much with my life. When I found that dream, I found hope. Hope not just for myself, but hope for us all. With that dream came a new way of looking at things. Bad things can be turned around and made to work for us.

I knew with that dream I needed courage. Courage to keep going and face the many obstacles to me realising it. To get that courage I drew on the inspiration of two very courageous women: Amelia Earhart and my mother. I used their courage to drive me, to lift me when I needed support, to find the courage in myself to endure till the end. I hope that in finding the courage that it took to complete my flight, and through writing this book, other people will gain the courage to pursue their dreams.

What after all is courage? The word itself is derived from the archaic French word cuer, which meant heart, and from which the modern French coeur — heart — is derived. Perhaps a truer meaning of cour-age is Age of the Heart and not the standard dictionary concept which most people seem to have, that it means to meet danger with fear. To me it means meeting danger with love.

I consider I have been courageous in my life, certainly during the preparation for my flight and throughout several incidents during it when my life was in danger. I know I have been courageous in writing this book. Each time I have been courageous I have been fearful. Have I been loving? Well, a cursory look by an outsider at my actions during such times would probably indicate that I haven't. Whether handling my plane when the engine stopped, or looking at the truth of my parenthood, curses, swear words, sobs or anger were my response and could be thought to be unloving, yet I know that I am more loving now than I have ever been in my life.

It has taken me a long time to learn, but thank God I have. Before I can love anyone, I must first love myself, and I do. I know in my heart that I am loving, and being loving puts my life on its true course.

Some people have described my flight as heroic. I certainly didn't live up to my own image of a heroine and it took me a long time to work out that heroes and heroines are ordinary people with similar fears and apprehensions to the rest of the world. I now have a better understanding that the two things that set them apart is their willingness to face their own fears and apprehensions, and the endurance to keep going until they reach their goal.

I was raised with the romantic idea that the heroic person was the one who did courageous things with a good deal of vitality and wit and handled the world with savoir faire. I certainly did not fit this image, and still don't. One of the great benefits to come out of the flight, for me, was the understanding that the fear and apprehension can provide the impetus to help any person to achieve their goals. I also came to understand that I am an ordinary person, capable of extraordinary things — like all of us.

I would like this book to be an inspiration to the many people who, like me, have reached that point in their life when they want to scream out 'What About Me!', and don't know where to turn. I say to them, take heart, help does come and usually in unexpected ways.

No one needs to fly an airplane around the world. That was for me to do. For me to prove my own worth to myself. Each person has their own challenges. Perhaps writing a book, perhaps applying for a mortgage, perhaps going to the local shopping centre. We each know in our hearts what it is we want to do for ourselves. With all my heart I encourage people to do it.

Do it with courage, from the heart, and you will succeed.

The final word on courage should be Amelia Earhart's that free spirit who lost her life in the pursuit of her dream.

Courage

Courage is the price that life exacts for granting peace.
The soul that knows it not, knows no release
From little things;

Knows not the livid loneliness of fear
Nor mountain heights, where bitter joy can hear
The sound of wings.

How can life grant us boon of living, compensate
For dull gray ugliness and pregnant hate
Unless we dare.

The soul's dominion? Each time we make a choice, we pay
With courage to behold resistless day
And count it fair.

Amelia Earhart

As my head hit the pillow on my first night home and I dozed off full of the excitement of the day, being home with my children safely tucked away in their beds in the same house and wondering what the future would bring, whether Neville and I would get back together again, or whether I would choose to stay a single mother, I'm not sure whether I really heard or dreamed it. No, I'm sure I heard it.

Good night, Gaby, and remember ... keep dreaming.

1	Oakland	**5**	Miami	**9**	Fortaleza	**13**	Gao
2	Burbank	**6**	San Juan	**10**	Natal	**14**	Fort Lamy
3	Tucson	**7**	Caripito	**11**	St Louis	**15**	El Fasher
4	New Orleans	**8**	Paramaribo	**12**	Dakar	**16**	Khartoum